Solid-State
ELECTRONICS
THEORY
with Experiments

M. J. Sanfilippo

TAB BOOKS Inc.
Blue Ridge Summit, PA 17214

This book is dedicated as a gift to my wife Molly and our son Michael. It is a small gift in comparison to the gifts of love, patience, and understanding that they have given me throughout our years together and especially during the writing of this book.

FIRST EDITION

FIRST PRINTING

Copyright © 1987 by TAB BOOKS Inc.

Printed in the United States of America

Library of Congress Cataloging in Publication Data

Sanfilippo, M. J.
    Solid-state electronics theory with experiments.

    Includes index.
    1. Solid state electronics. 2. Solid state
electronics—Experiments. I. Title.
TK7871.85.S33 1987   621.381   87-7103
ISBN 0-8306-0926-1
ISBN 0-8306-2926-2 (pbk.)

Questions regarding the content of this book
should be addressed to:

    Reader Inquiry Branch
    Editorial Department
    TAB BOOKS Inc.
    P.O. Box 40
    Blue Ridge Summit, PA 17214

Cover photograph courtesy of John Sedor Photography.

Contents

Introduction

The study of solid-state electronics concentrates on the practical applications of small electronic devices called diodes and transistors. Although many technicians and engineers attribute the invention of the transistor as the stepping stone into today's modern electronic era, it was actually the introduction of the simple junction diode that revolutionized the world into the electronic age that we live in today. From the invention of the first crystal radio, with a single crystal and diode, to the foundation of digital electronics, with integrated circuits made up of a seemingly infinite number of diodes called gates, the junction diode has played an important role in the way in which we interact with our world today.

In retrospect, the transistor, an extension of the diode concept, opened a new vista in the world of electronics. Where a large, high current, high voltage vacuum tube was needed to meet specific demands, the smaller, lighter weight, less power consuming transistor stepped in to fill those same needs. And in many cases the transistor costs far less to purchase and operate than the vacuum tube.

From the basic two-junction transistor come variations of solid-state devices such as SCRs, triacs, and diacs not to mention field effect transistors (FETs) and opto-isolators. Even the first integrated circuits (ICs) were manufactured using similar processes that have been used in the manufacture of transistors. These discrete devices form the basic building blocks of many of today's electronic circuits and even complete systems, from power supplies to power amplifiers. In these pages you will discover how these solid-state devices operate, and learn about their various applications.

Although this book has been written so that it may be read as a self-study, it can also be used effectively in the classroom setting. Performance objectives, those objectives that measure the student's hands-on ability in electronics, are usually left to the lab manual and to laboratory exercises. However, there are circuits at the end of the text, giving a practical aspect to the material presented, that can be constructed and perhaps it is here that the instructor may wish to evaluate the student on the basis of performance. I would suggest, however, that this type of evaluation be better left to laboratory experiments and that, instead, the circuit construction projects at the end of the text be used to reinforce the concepts of the material presented. Experience dictates that if the student can successfully complete a project that includes concepts just discussed, then that student has more than met the cognitive objectives established by the instructor.

In addition, mathematics has been kept to a minimum. This is not to disqualify the technical merit of the material in this study, but to prevent the reader from getting lost in a myriad of formulas and equations. In most cases higher mathematics simply is not used when working as a technician in the field of solid-state electronics. Heavy mathematics, with an emphasis on calculus, is more useful to the design engineering oriented person than to the technician who must understand the concepts behind an inoperative circuit and then repair that circuit to a functional level.

The outline of the material here follows that suggested by the American Electronics Association in its guidebook "Curriculum Requirements For Electronic Technician Training" published in November of 1980. I am grateful to that taskforce for the amount of effort they put into the research to establish those guidelines

It is therefore the intent of this book to enable a student in solid-state electronics to grasp concepts in this field of study, and hopefully to enhance those concepts with some optional hands-on experimentation. It is also the hope of this author that the instructor will inject into the classroom setting the experience gained working with solid-state devices and the enthusiasm needed to prepare for a new venture in the study of solid-state devices.

Chapter 1

Introduction to Semiconductors

Semiconductor materials are the basic building blocks used to construct electronic components or devices such as diodes, transistors, and integrated circuits. The simplest of these devices is the single junction, or unijunction diode. With some variation, other electronic devices such as transistors and integrated circuits are constructed based on the design of this diode. In turn, complete electronic circuits can be designed using a combination of all three of these solid-state devices.

Essentially, the purpose of any semiconductor device is to control the flow of current or to control the amount of voltage within a given electronic circuit. In this sense, semiconductors are actively involved in the control of electrons and are considered active devices. Other components such as resistors, coils, and capacitors, are known as passive devices because they cannot vary electron flow, only store it or oppose it in some way.

There are, of course, advantages and disadvantages to the use of semiconductors, or solid-state devices, in electronic circuits and equipment. Advantages are discussed first to see what has made semiconductor devices the prevailing electronic component in use today.

ADVANTAGES OF SEMICONDUCTORS

The terms semiconductor and solid-state device are often used interchangeably because most, but not all, semiconductor materials

1

are made from solid materials. This leads us to the first advantage of semiconductor devices; that is, their reliability due to sturdy construction. They are not prone to breakage as the glass envelope of a vacuum tube is if it is bumped or dropped.

Also, additional power is not needed to heat filaments or heaters that are found in vacuum tubes. Those same filaments are somewhat limited in life expectancy and require a warm-up period before operating properly. In contrast, the semiconductor device begins operating at its designed parameters almost immediately after power is applied and therefore requires virtually no warm-up time.

Solid-state devices require less power than vacuum tubes because they usually operate at lower voltages and currents. This also makes them safer to work with and allows the equipment designed with these devices to be battery operated and portable. And because of their small size they are less expensive to produce and buy than the vacuum tube.

Finally, by using a special manufacturing process, a great many devices can be combined into a very small solid package called an integrated circuit, or IC. ICs may contain hundreds of transistors and diodes making up a complete single circuit or even an electronic system made up of dozens of circuits. The advantages then are numerous, and great strides in circuit and equipment development have taken place due to the introduction of solid-state devices.

DISADVANTAGES OF SEMICONDUCTORS

With all of the advantages stated above it may be difficult to imagine that there are disadvantages to solid-state devices. However, they do exist, although they are rapidly declining in number. This is due in part to new technology in the manufacturing processes of these devices, allowing them to become available for almost all of the needs and applications of the design engineer and technician.

Probably the biggest disadvantage of solid-state devices is their inability to operate consistently over a wide range of temperatures. To compensate for this, additional circuitry is usually needed to maintain the design or operational parameters of the device. In some instances these devices are irreversibly damaged if operated above or below their stated temperature rating for an extended period of time. They may also be damaged if the operating voltages applied to them are accidently reversed.

Finally, and this is changing quite rapidly, many solid-state devices are not able to operate at high power and high frequency at the same time. There are high power, low frequency devices and low power, high frequency devices, but few devices that are able to operate at both high power and high frequency simultaneously. One Japanese company does manufacture a solid-state device using gallium arsenide called a GaAsFET (pronounced gass fet) that operates in the gigahertz (1×10^{12}) frequency range with a power of up to ten watts, but at this writing there are still a few problems in characterizing or matching these devices with the circuit in which they are operated. These types of devices are discussed in more detail in a later chapter.

In almost all aspects of your work in electronics you will be working with solid-state devices. In just a few short years semiconductors have become the mainstay of electronic circuits. As new manufacturing processes develop, new ways in which to use solid-state devices will also unfold allowing us to enhance the methods in which we do things to an even greater extent than ever before thought possible.

SEMICONDUCTOR MATERIAL

The term semiconductor is used to describe a device whose properties allow it to be classified as neither a conductor nor an insulator but rather as a device whose properties or characteristics lie somewhere in between. It is not as effective as an insulator in preventing current from flowing through it, and yet it does not allow current to flow through it as easily as a conductor does.

Germanium and Silicon

The materials most often used in the manufacture of semiconductors are germanium and silicon. Germanium is a rather brittle, grayish-white element found in the earth while silicon is a non-metallic element found in the earth's crust. The white sand found on many beaches is actually a variation of silicon and is called silica. Through a chemical process, pure silicon can be obtained from silica. Both germanium and silicon have atomic structures allowing them to be chemically altered for suitable use as semiconductors.

Semiconductor Atoms and Crystals

The atomic structure of semiconductor materials must first be examined before a study of solid-state devices can be made, because the simplest of devices upon which most other solid-state devices are designed, the unijunction diode, can best be explained by viewing the physical structure of semiconductor material itself. An in-depth study in physics is not required, but a simple overview of the physical nature of semiconductor materials is helpful. Much of this material is similar in content to the basic atomic theory used to explain the nature of electron flow through a conductor that you may have read about in your introductory studies of basic electronics.

The atom is composed of three basic parts or particles, protons, neutrons, and electrons. The center of the atom is called the nucleus and contains the protons and neutrons. Surrounding, or orbiting, this nucleus are the electrons. Figure 1-1 is an illustration of this concept.

This configuration of the atom is set up much the same way that our solar system is, with the planets orbiting the sun. However, the electrons are not orbiting the nucleus in a flat plane, but rather in various planes and in various distances from the nucleus, or in layers called shells. The outermost shell is called the valence shell and can contain a maximum of eight electrons. It is the amount of electrons in the valence shell that determines whether an element made up of these atoms is an insulator, a conductor, or a semiconductor.

An element with one to three electrons in the valence shells of its atoms is an electrical conductor. If there are exactly four electrons in the valence shells of the atoms making up the element, then it is a semiconductor. And if there are five to eight electrons in the valence shells, then the element is an insulator.

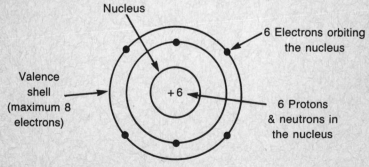

Fig. 1-1. A typical carbon atom.

Elements with five or more electrons in their outer shell try to become stable by acquiring enough electrons until a maximum number of eight electrons are orbiting about the nucleus in the valence shell. Because of this tendency towards stability, these elements make good insulators. Insulators contain atoms that try to acquire electrons rather than allowing them to drift randomly. In contrast, those elements with less than four electrons in their valence shells allow those electrons to be given up easily and therefore are good conductors. Remember, it is the movement of these random electrons that supports current flow. Elements made up of atoms that contain four electrons in their valence shells make neither good insulators nor good conductors. These elements make up the materials called semiconductors.

How the atoms of a semiconductor such as silicon are held together is discussed next. Figure 1-2 shows silicon atoms in a crystal structure. The atoms are held together by a process called covalent bonding.

Remember that the outermost shell of an atom is called its valence shell. Covalent bonding simply means that the electrons in the valence shell of one atom are allowed, under certain conditions, to share the valence shell of the adjacent atoms. In essence, this sharing of electrons lets us view each atom as having a total of eight electrons in its outer shell rather than the original four. This structure, called a crystalline structure, now becomes electrically stable, very much like an insulator, since there are now eight electrons in the valence shells.

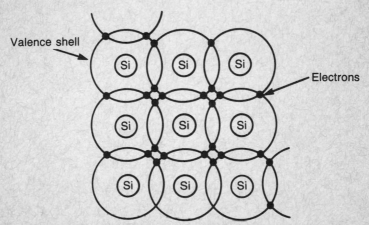

Fig. 1-2. Silicon atoms in a crystal structure held together by covalent bonding.

The atomic structure of a silicon semiconductor is shown in Fig. 1-3. Notice that this semiconductor material has only three shells. The innermost, or first, shell contains two electrons. The second shell contains eight electrons, and the outermost shell, or valence shell, contains four electrons. There are therefore a total of fourteen negatively charged electrons.

The nucleus is at the center of the atom. It contains an equal number of positively charged particles called protons. There are fourteen electrons and fourteen protons in the silicon atom—in fact, there are always an equal number of protons and electrons in all atoms.

The crystal structure described earlier is sometimes also referred to as a lattice structure, or crystal lattice. The crystal lattice of the semiconductor material of silicon shown in Fig. 1-2 is referred to as an intrinsic material. It does not contain atoms other than silicon or any other impurities, and is therefore a very pure semiconductor material.

CONDUCTION IN INTRINSIC MATERIAL

Temperature plays an important role in the conduction of current in a semiconductor material. As the ambient temperature surrounding an intrinsic material is lowered, the valence electrons of these materials tend to cling to their parent atoms rather than to drift randomly at times. At extremely low temperatures these valence electrons remain tightly bound to their atoms and the element made up of these atoms tend to act very much like insulators rather than semiconductors. Since current flow depends

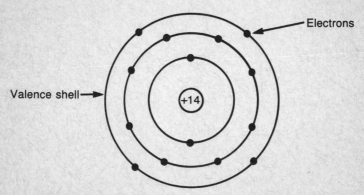

Fig. 1-3. A simplified diagram of the silicon atom with 4 electrons in the valence shell.

on the movement of electrons through a semiconductor, the material does not allow current to flow and the material therefore acts to prohibit the flow of current.

Just the opposite holds true at extremely high temperatures. At that end of the temperature spectrum valence electrons become quite active and a few break away from their covalent shells to drift from one atom to another. These are called free electrons. If a voltage source is applied to the semiconductor material at these higher temperatures, these free electrons would be able to support current flow. As you might expect, the higher the temperature, the more the semiconductor behaves as a conductor. However, semiconductors are not operated at these high temperatures. Instead, another method is employed to allow semiconductors to more efficiently conduct current. This requires looking a little more closely at what is taking place inside the semiconductor material itself.

Hole Flow

Whenever an electron breaks away from a covalent bond, a vacancy is left in that valence shell. That vacancy or space is referred to as a hole. Since an electron has a negative charge, the absence of an electron can be viewed as having the opposite charge, or it can be said that the hole created by the absence of the electron carries a positive charge. In effect, the movement of the electron from one atom to another creates a positively charged particle called a hole. When an electron moves to the valence shell of another atom to create a hole in the valence shell of its original atom, an electron-hole pair has been created as well as the positively charged area or particle that has been referred to as a hole. This is illustrated in Fig. 1-4.

Again, temperature does have an effect on the number of electron-hole pairs found in semiconductor material. As the temperature increases, more electron-hole pairs are formed. In the case of these small numbers of pairs found even at room temperature, electrons tend to drift randomly but are, in most cases, more readily absorbed by the holes. This inhibits the movement of electrons and limits them to random jumps, perhaps only to the next atom. When that happens, a hole is created, and in the process, it appears that the hole is moving in the opposite direction from the electron. Holes move in the direction opposite to that of the electron.

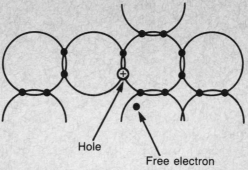

Hole

Free electron

Fig. 1-4. The movement of an electron through a semiconductor material creates a hole.

Another view of this concept is to picture an electron moving from one atom to another. Each time the electron moves, it creates a hole, which is now in the location the electron was in. When another electron fills that first hole, that first hole now is in the spot where the second electron was. What can be seen here is that the hole is moving in a random manner through the semiconductor material just as the electrons are moving randomly, but instead are moving in the opposite direction.

Current Flow

An intrinsic material such as silicon can have a difference of potential applied across it as shown in Fig. 1-5.

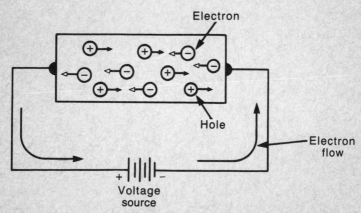

Fig. 1-5. Current flow in an intrinsic semiconductor material consists of both electron flow and hole flow, each moving in the opposite direction towards the voltage source.

In this instance, the free electrons labeled with a minus sign in a small circle are attracted to the positive side of the voltage source. Remember, like charges repel, unlike charges attract. In the same way that the free electrons are attracted to the positive side of the voltage source, the holes created by the moving free electrons are attracted to the negative side of the voltage source. In this illustration, for simplicity, electrons are moving from left to right while the holes are moving from right to left rather than randomly. As the positive side of the voltage source receives the free electrons, an equal number of electrons are leaving the voltage source and flowing into the left side of the semiconductor material. Some of the holes in the material naturally absorb some of these new electrons arriving from the voltage source. The holes are therefore always moving to the left and the electrons are always moving to the right as long as the source voltage is connected in this manner.

The biggest difference between a conductor and a semiconductor in respect to current flow is that:

- In a conductor, you can think of the flow of electrons only as supporting current flow.
- In a semiconductor, you can think of the flow of both electrons and holes as a prerequisite to the flow of current.

Remember that in a semiconductor, current flow consists of both hole flow and electron flow. Holes are considered to be positively charged particles since their presence is an opposite and obvious result of electron movement. Electrons are negatively charged particles and move in a direction opposite to that of the holes that they themselves create as a result of their movement. Also keep in mind that as ambient temperature increases, electron-hole pairs increase in number and therefore the capability of a semiconductor to conduct current also increases.

CONDUCTION IN DOPED SILICON

You now know that current flow in a semiconductor consists of both hole flow and electron flow. However, there is a method of increasing the conductivity of a semiconductor without raising the temperature to a point where the semiconductor may be damaged. This is accomplished by adding an impurity to the semiconductor material, called doping. These impurities are usually added to the material when it is first produced. This discussion of the doping process is limited to the semiconductor material silicon primarily because the same process of adding an impurity to increase conductivity is also applied to other semiconductor materials such as germanium.

There are only two types of impurities added to intrinsic materials to increase their conductivity. The first is a pentavalent substance which is made up of atoms that contain five electrons in their valence shells. This means, in effect, that extra electrons are being added to the already pure material that contains exactly four electrons in the valence shells of its atoms. The second type of additive is termed a trivalent material. The atoms that make up this material contain only three electrons in their valence shells, so holes are being added to the intrinsic material that is being readied for better conductivity.

N-Type Semiconductor Material

The process of doping a semiconductor material means adding an impurity such as arsenic during the production process. The addition of this particular kind of dopant adds electrons to the overall crystalline structure of the silicon since arsenic is a pentavalent element. Also, this additional element replaces some of the silicon atoms with atoms that are referred to as donor atoms since the arsenic atom has an extra electron in its valence shell. Figure 1-6 shows the crystalline structure of the doped element silicon with the addition of the impurity, or arsenic atom.

The arsenic atom behaves very much like the silicon atom in creating a covalent bond and in the sharing of its electrons. Notice, however, that there is an extra electron in the valence shell of the

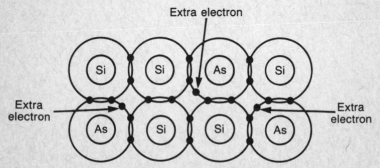

Fig. 1-6. Pentavalent doping of a silicon crystal structure creates an N-type semiconductor material when arsenic is used in the doping process.

arsenic atom. This is the donor atom, which is rather loosely attached to its parent atom and with a little help from a voltage source can be set free to jump to another atom. Remember that there are now a great many of these donor atoms within the semiconductor material, so there are now a greater number of free electrons as well. An intrinsic material like silicon that has been doped with a pentavalent element and that now has many additional electrons is also more negatively charged than it was originally. To distinguish this material from another type of semiconductor material discussed soon, this new material is called an N-type semiconductor. N is an abbreviation for the word negative, since electrons are negatively charged particles.

Earlier, a voltage source was applied to an intrinsic semiconductor material. If that same voltage source is now applied to a semiconductor material that has been doped with a pentavalent element, a somewhat different, but similar, type of current flow takes place. In this case, there is a considerably greater flow of electrons because of the added free electrons from the impurity, shown in Fig. 1-7.

The donor electrons are flowing towards the positive side of the voltage source. Along with these donor electrons are the original electrons that broke away in the intrinsic semiconductor material that formed the electron-hole pairs that were mentioned earlier. The holes from the electron-hole pairs begin to move towards the negative side of the voltage source while the electrons from these electron-hole pairs and the electrons from the donor atoms head in the direction of the positive side of the voltage source.

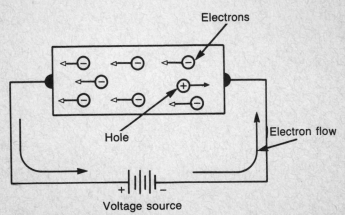

Fig. 1-7. A voltage potential applied across an N-type semiconductor material.

The electrons in the doped semiconductor material now greatly outnumber the holes. In effect, it can be said that they are in the majority. Therefore, they are called majority carriers. It would follow then that the holes are referred to as the minority carriers.

P-Type Semiconductor Material

The introduction of a trivalent such as gallium to an intrinsic semiconductor material produces the crystalline structure shown in Fig. 1-8. Since the gallium atom only has three electrons in its valence shell it cannot share a fourth atom and therefore the missing electron is actually considered a hole, or a positively charged particle. Since this intrinsic material has now been doped with a trivalent, a large number of holes are free to drift within the semiconductor. There are now a great many more holes than electrons. And, as you may have guessed, it is now referred to as a P-type semiconductor. In this case the P simply means positive.

An atom with an additional electron in its valence shell is referred to as a donor atom because it donates an extra electron to the covalent bond among the adjacent atoms. In the case of the P-type semiconductor material, the atoms that have a fourth electron missing from their valence shells are called acceptor atoms because these atoms readily accept the electrons that are a part of the electron-hole pairs that are drifting throughout the semiconductor material.

In the same way that a voltage source was applied to the N-type semiconductor, a voltage source can also be applied to a P-type semiconductor with similar results, illustrated in Fig. 1-9. In

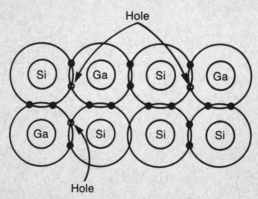

Fig. 1-8. Trivalent doping of a silicon crystal structure creates a P-type semiconductor when gallium or aluminum is used in the doping process.

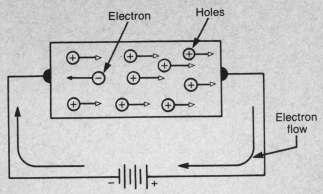

Fig. 1-9. A voltage potential applied across a P-type semiconductor material.

this case the movement of holes towards the negative side of the voltage source is the same as in the N-type semiconductor, and the movement of electrons towards the positive side of the voltage source is also the same as in the N-type semiconductor. However, because of the addition of the acceptor atoms, there are many more holes than electrons.

There are also electron-hole pairs that are formed as electrons break away from their silicon atoms. These holes are in addition to those provided by the acceptor atoms and thus in this type of semiconductor the holes are in the majority, making them the majority carriers while this time it is the electrons that are the minority carriers.

It is important to remember that although the N-type material has more electrons it is not negatively charged, nor is the P-type semiconductor material positively charged simply because it contains more holes than electrons. This is because each atom within each type of semiconductor still contains an equal number of electrons and protons and therefore remains electrically neutral. Also keep in mind that as more dopant is added to the semiconductor material the greater is its conductivity as compared to the conductivity of pure semiconductor material.

The doping of semiconductors into N-type and P-type may seem quite useless to you unless it can be made, in some way, to work for us. That very concept is explored in the next chapter on semiconductor diodes.

Chapter 2

Semiconductor Diodes

To better understand other solid-state devices such as transistors, thyristors, and integrated circuits, it is best to first grasp an understanding of the most basic of solid-state devices, the PN junction diode. This is because these other devices are constructed and operated using the same principles as the simple diode.

PN JUNCTION DIODES

The previous chapter discussed how a semiconductor material can be made more conductive by the addition of an impurity such as arsenic or gallium. The addition of these impurities adds to the flow of electrons or holes depending on whether the intrinsic semiconductor material has been doped as an N-type or P-type semiconductor. Remember, in a semiconductor, current flow is dependent on both electrons and holes, not on electrons only. Also, both types of semiconductors are still electrically neutral and do not possess a negative or positive charge. This is because both the semiconductor atoms and the impurity atoms contribute the same number of electrons and protons in their respective materials.

Another way of stating this is to say that the impurity atoms and the intrinsic atoms are neutral and remain so in spite of their covalent bonding, so the overall doped semiconductor remains neutral in charge. However, independent electrical charges do exist within each type of semiconductor material and this is the topic of our next discussion on how the PN junction diode is able to control current flow.

Positive and Negative Ions

Although the overall charge of an N-type or P-type semiconductor material remains electrically neutral, each time an atom gives up an electron within the semiconductor, it loses a negative charge and is no longer electrically neutral. The atom now possesses a positive charge because the protons outnumber the electrons. Since the atom has taken on a somewhat different form electrically, it is now called an ion. An atom that gives up an electron is called a positive ion, while an atom that accepts an electron is called a negative ion.

The acceptor and donor atom (described in Chapter 1) that make up the impurity elements are those same negative and positive ions that are now being referred to, respectively. Although the atoms are now either positive or negative ions, since they have given up or accepted electrons, the holes and free electrons drifting throughout the semiconductor material also possess positive and negative charges, respectively. These are called mobile charges and are equal in number to the ionic charges. Once again these two types of charges are equal in strength and opposite in charge and therefore the N-type or P-type semiconductor material remains electrically neutral. It's important to understand doped semiconductors as containing positive or negative ions rather than donor or acceptor atoms so that you can better understand current flow within and through a PN junction diode. Figure 2-1 illustrates this concept.

In this figure you can see that the N-type semiconductor has been doped with a pentavalent impurity. These pentavalent atoms are represented by a circle with a plus sign since this donor atom gives up an electron and becomes a positive ion. It also donates a free electron, represented by the minus sign next to the symbol for the positive ion. In a similar manner, the P-type semiconductor has been doped with a trivalent impurity, shown as a circle with a minus sign inside of it. That is because this atom accepts an extra electron and therefore is a negative ion. These atoms create holes which are represented as plus signs next to each negative ion. The next step in producing a semiconductor diode is to physically join together the N-type and P-type semiconductors.

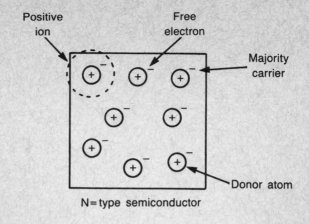

N= type semiconductor

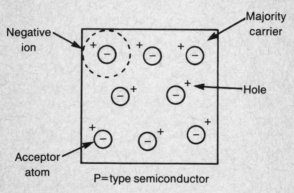

P=type semiconductor

Fig. 2-1. Ions are formed within an N-type and P-type semiconductor by the addition of donor atoms or acceptor atoms, respectively.

The Depletion Region

The area where the two types of semiconductors meet, the PN junction, is where the diode is able to control current flow. Figure 2-2 shows the PN junction.

Initially, electrons in the N-type semiconductor are attracted to holes in the P-type semiconductor. In the same way, holes in the P-type semiconductor are attracted to the electrons in the N-type semiconductor. As the free electrons cross the junction, the N-type material becomes depleted of electrons near this junction. The same thing happens to the holes in the P-type semiconductor. An area along the junction known as the depletion region has been created. This region is created very quickly and is relatively small in size along the junction, because the electrons that have crossed

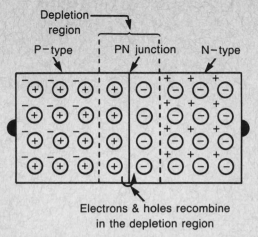

Fig. 2-2. When N- and P-type semiconductors are joined together a PN junction is formed, creating a depletion region.

over to the P side serve as a negative potential to repel any further electrons from crossing over. The holes that moved over into the N side now also repel any other holes from crossing over. Keep in mind that the depletion region is now free of majority carriers containing ionized atoms. The area on the N side of the junction now has a positive charge, while the area the P side of the junction contains atoms that have lost holes and therefore has a negative charge. These charges exist only in the junction, while overall the PN junction diode remains electrically neutral.

Barrier Voltage

In addition to the depletion region, a potential difference or voltage known as a barrier voltage has now been created, shown in Fig. 2-3.

The barrier voltage or barrier potential that exists across the junction can just as easily be represented by an external battery. This voltage varies from semiconductor material to material. It is generally about 0.7 volts for silicon semiconductor diodes and about 0.3 volts for a germanium semiconductor diode. Remember that this barrier voltage exists within the semiconductor itself and cannot be measured directly; however, it plays an important role in the operation of the diode.

The symbol for a diode is shown in Fig. 2-3. Electrons flow in the direction opposite to the direction in which the arrow is

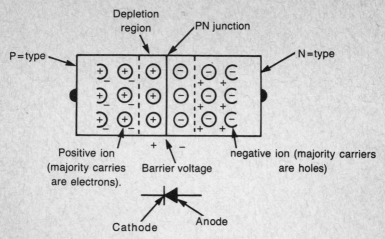

Fig. 2-3. Barrier voltage is produced when electrons in the N-type semiconductor and holes in the P-type semiconductor combine at the junction.

pointing. (I use electron flow rather than conventional current flow in dealing with the operation of semiconductors in the pages to follow.)

MANUFACTURING TECHNIQUES

Now that you have a basic understanding of the electrical construction of the PN junction diode it may be of some interest to see how the diode is constructed physically. From there you will take a look at the action inside of the diode when applying voltages in the forward and reverse directions.

Several methods are employed in the manufacture of semiconductors. The actual process begins with obtaining a pure intrinsic material to form the semiconductor material that is needed to eventually become the N-type and P-type semiconductors that have been discussed.

The first step is to obtain a pure monocrystalline material that can be doped to become either an N-type semiconductor or a P-type semiconductor. Through chemical processing of silica into silicon, a polycrystalline silicon material is formed; however, what is needed is a monocrystalline silicon material. The final silicon is obtained by placing the polycrystalline silicon into a crucible. A monocrystalline material has an orderly structure or continuous arrangement of atoms that form the crystal. The polycrystalline material is placed into a crucible and melted down into a liquid to prepare it to become the structured crystal that is needed. It is during this melting process that the impurity is added to make the polycrystalline a P-type or N-type semiconductor material. Figure 2-4 shows the process in a somewhat simplified form.

Here, the seed crystal, a small bit of monocrystalline silicon, is lowered into the crucible until it just touches the surface of the melted silicon. This melted silicon begins to crystallize onto the seed as a monocrystalline structure, since the seed crystal is much cooler than the melted silicon. As soon as this begins to happen, the seed crystal is pulled slowly upward while more and more of the silicon accumulates, forming a larger and larger crystal. Eventually, the crystal is formed in a cylindrical shape approximately two to four inches in diameter with a length of ten to twelve inches. For continuity of the crystal growing process, the crucible and seed are slowly rotated in directions opposite each other during this process.

Now the cylinder is cut into thin slices called wafers. Each wafer is then highly polished and then finally cut into tiny chips, usually by an extremely precise laser beam, to form the foundation for the PN junction diode. In this way, literally thousands of diodes can be manufactured from one cylinder of semiconductor material. The next step is the making of the diode junction.

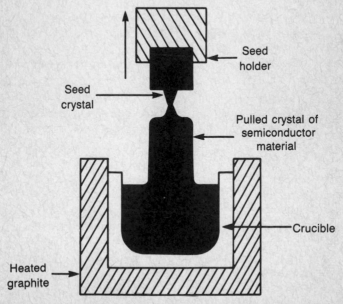

Fig. 2-4. Simplified diagram showing a pure semiconductor crystal being produced by a seed crystal.

19

Today, some diode junctions are formed using a method known as epitaxial growth where one type of semiconductor is grown onto another, forming a junction where the two semiconductors meet. Figure 2-5 gives a simple rendition to the concept behind epitaxial growing.

Each of these wafers is placed on a heated surface inside of a chamber that is fitted to allow the addition of a gas such as silicon chloride. This gas also contains a very small amount of boron, a trivalent element. As the gas comes into contact with the heated silicon wafer, it decomposes, leaving a P-type monocrystalline semiconductor coating the layer of N-type semiconductor wafer. Where the two semiconductors come into contact with each other, a PN junction is formed.

PACKAGING

The PN junction diode must now be placed into some kind of package that makes it easy to use in an electronics circuit. A number of methods are used depending upon the way in which the diode was manufactured. One of the simpler methods is just to place the diode between two conductive materials, and attach wires to those materials. It's not quite that easy. Figure 2-6 gives you an idea of the method that is used to mount a diode so that it is of some practical use to us in our work with solid-state devices.

This illustration is called a cross section because if a section of this diode was cut across what would be seen is what is shown here. The diode chip, sometimes referred to as a pellet, is shown sandwiched between two conductive materials. On top, an electrical connection is made to the top of the diode pellet, which is the P-type material. This is called the anode portion of the diode. The

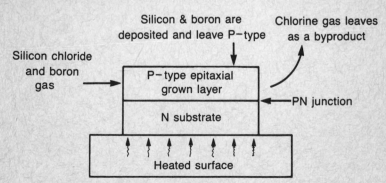

Fig. 2-5. Simplified version of epitaxially grown PN junction diode.

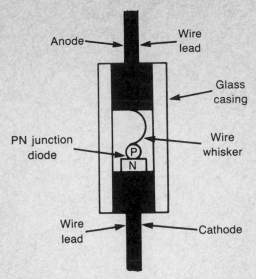

Fig. 2-6. Mounting a PN junction diode in a case with leads allows for its use in electronic circuits.

bottom of the pellet is soldered to the bottom of the case, or package, and it is this case, in which the diode is mounted, that forms the cathode. The cathode is always the negative side of the diode, while the anode is the positive side of the diode. When this diode is connected in an electronics circuit to perform a function, electrons enter the cathode and leave at the anode.

There are, of course, many other types of diode packages. Some are very small and made from glass, with the diode enclosed within. Some are quite large and are made of metal. The size of the diode is usually a good indication of the amount of current that it is able to carry or pass. In all cases, the diodes are hermetically sealed, meaning that they are airtight. Figure 2-7 shows some types of diode packages used in electronic circuits.

Fig. 2-7. Some of the packages in which diodes are mounted—different sizes usually indicate current and/or power ratings (John Sedor Photography).

BIASING

The diode is the first semiconductor or solid-state device that will be used to control current flow. To accomplish this it is necessary to apply voltages to the diode that allow it to perform this function. The voltages that are applied to the diode are called bias voltages and control the operation of the diode in exactly the manner for which it is designed.

Bias voltages allow the diode to conduct or prevent the diode from conducting current flow altogether. The type of bias voltage that allows current flow through the diode is called forward bias. That voltage that prevents a diode from conducting current flow is referred to as reverse bias. Both types of bias play an important role in how a diode operates and both types are necessary in the proper operation of other solid-state devices as well, such as transistors and integrated circuits. A discussion of each of these bias voltages, beginning with forward bias, shows how this is the way in which most diodes are biased in order to perform properly. Not all diodes are biased in this manner. Some operate properly in the reversed biased mode, such as the zener diode. This diode and other special purpose diodes are discussed later on in this chapter.

Forward Biasing

Free of any external or outside voltage sources, the PN junction diode contains an inherent voltage potential across its junction of approximately 0.7 volts for silicon and 0.3 volts for germanium. This is because the majority carriers combine at the junction to form a depletion region. Although free from these carriers, this region contains positive and negative ions which are atoms that contain positive and negative charges. It is the buildup of these charges within the depletion region that creates the difference of potential or barrier voltage.

Connecting an external voltage source to a PN junction diode has an interesting effect on the depletion region and barrier voltage, particularly when the external voltage is significantly higher than the barrier voltage. Remember that this theory applies to both silicon and germanium diodes as well. Figure 2-8 is an example of what takes place within the diode when a battery is connected to each side of the diode.

In this illustration, the negative side of the voltage source is connected to the N-type semiconductor side of the PN junction diode while the positive side of the voltage source is connected to the P-type semiconductor side of the diode. The free electrons in the N side are repelled by the negative side of the voltage source and travel towards the junction where they combine with the positive ions in the depletion region. The positive ions now become electrically neutral once again. At the same time the negative ions on the P side of the diode are attracted to the positive side of the voltage source. This in turn causes the negative charge on the P side of the diode to become electrically neutral. You can now see that without a negative and positive charge on either side of the PN junction, there is effectively, no barrier voltage. Another way of saying this is that by placing an external voltage on the diode we can neutralize the barrier voltage which previously prevented the flow of current across the PN junction. The diode now supports a continuous flow of current.

As shown in Fig. 2-8, electrons from the voltage source flow into the N side of the diode. The free electrons in that side of the diode support this flow of electrons towards the junction. The same thing happens on the P side of the diode. The holes, or majority carriers, on that side flow to the junction where they combine with the incoming electrons. Now the holes and electrons, in effect, neutralize each other. But electrons and holes are still appearing along the outer edges of the diode, provided by the external voltage source, and continue to move towards the junction and to combine. Electrons flow through the entire PN junction diode from the N side to the P side and back to the positive side of the voltage

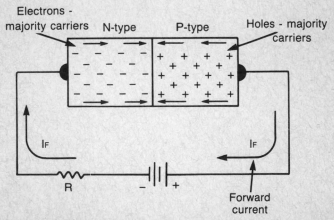

Fig. 2-8. A forward biased PN junction diode with resistor R used to limit current flow to a safe value.

source. Since the electrons are moving from the negative side of the external source, through the diode, and back to the positive side of the voltage source, the diode is said to be forward biased. Since the internal diode resistance is usually very low it is necessary to limit the large flow of electrons through the diode by the use of the external resistor as shown in Fig. 2-8.

It is safe to say that any diode conducts current provided that the external voltage source is sufficiently high enough to overcome the diode's barrier voltage. For silicon diodes this voltage is approximately 0.7 volts and for germanium diodes this voltage is approximately 0.3 volts. Remember that the polarity of the external voltage source must be connected correctly for conduction of the diode. This means that the negative side of the voltage source must be connected to the N side of the diode, or to its cathode, and the positive side of the voltage source must be connected to the P side of the diode, or to its anode. The external voltage needed to overcome the barrier voltage is termed the diode's forward voltage drop, or V_F. Since an external resistor has been added to the bias circuit, the resistance of which is already known, and also the V_F of the diode is known, you can now calculate the current flow through the diode. It's simply a matter of using Ohm's law and putting the correct numbers into the formula. One other value must be known and that is the value of the dc bias voltage.

Assume that a silicon diode is connected to a battery of 10 volts dc with an external series resistor of 1000 ohms. To find the amount of forward bias current, or I_F, the following formula is used:

$$I_F = \frac{V - V_F}{R}$$

I_F = Forward current
V_F = Forward voltage
V = dc bias voltage
R = Series limiting resistor

Substituting the given values in the above formula:

$$I_F = \frac{10 - 0.7}{1000} = .0093 \text{ amps or } 9.3 \text{ mA}$$

24

Reverse Biasing

Further examination of a PN junction diode shows that if the battery connections of Fig. 2-7 are reversed, the diode does not conduct current flow. By forward biasing the diode current is allowed to flow. By reverse biasing the diode, current is prevented from flowing through the circuit. In effect then, the diode acts very much like a switch. However, a small current does flow even though the diode is reverse biased. This is shown on Fig. 2-9.

The diode in this figure is reverse biased because the negative side of the battery is connected to the P side of the diode and the positive side of the battery is connected to the N side of the diode. The depletion region becomes considerably wider or larger along the junction within the diode, because the positive side of the battery attracts the free electrons in the N side of the diode, pulling them away from the edge of the depletion region making it even wider on the N side. Thus, the positive charge on this side of the depletion region will increase. While this is taking place, electrons are leaving the negative side of the battery and entering the P side of the diode. They begin to fill the holes near the junction. Since holes move in the opposite direction of electrons, they appear to be moving in the direction of the positive side of the battery. This action actually creates negative ions near the junction of the P side of the diode, widening the depletion region on that side. What has happened is that by reverse biasing the diode the depletion region has been increased to a point where the diode simply does not conduct current flow. Remember, eliminating the depletion region and thus the barrier voltage allowed the diode to support the flow of electrons and holes necessary for current flow. Therefore, widening this re-

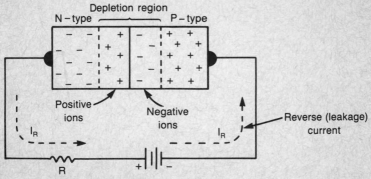

Fig. 2-9. A reverse biased PN junction diode.

gion has the opposite effect, preventing the flow of current. Also, by reverse biasing the diode, the barrier voltage builds up to the voltage of the external bias source.

Up to this point majority current carriers flowing through the PN junction diode have been discussed. However, even though the diode is reverse biased, a very small current does flow. This is called leakage current and is supported by the minority carriers within the diode. On the N side of the diode the minority carriers are holes and on the P side of the diode the minority carriers are electrons. These minority carriers are forced towards the junction. It is here that they combine and support a very small current flow. As temperature increases, more electron-hole pairs are created and leakage current also increases. Although this leakage current is usually only in the microampere range, it is considered important in some applications. Manufacturers of solid-state devices usually specify leakage current in their engineering application notes and these are used by the technician and engineer in designing certain electronic circuits.

CHARACTERISTIC CURVES

One of the easiest ways to understand how a diode operates electrically is by the use of a characteristic curve. This is used extensively in explaining the operation of other devices too, such as transistors. It is similar, in fact, to the curves used to represent capacitor charge and discharge times that you may have read about in your previous studies of basic electronics. A graph shows how voltage, current, and even how temperature are all related during the operation of a diode.

Figure 2-10 is an illustration of a characteristic curve of a typical silicon PN junction diode. Although both silicon and germanium diodes have similar characteristic curves, there are some distinct differences between the two. These will be examined shortly. The diode's forward current, I_F, is plotted on the vertical axis above the horizontal axis. The bottom half of the vertical axis contains information on the diode's reverse current, I_R. This is the leakage current. Notice that it remains very small, almost insignificant, until a certain reverse voltage, V_R, is reached. Once this particular point is reached, the diode breaks down. It is at this time that a high reverse current begins to flow. The reverse voltage at that point is appropriately called the breakdown voltage. This voltage

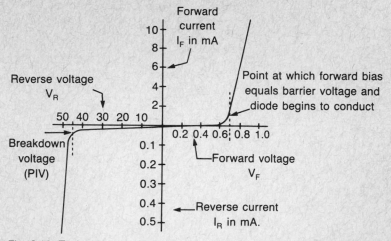

Fig. 2-10. Typical V-I characteristic curve of the silicon PN junction diode.

varies from diode to diode since no two diodes are constructed exactly alike.

Many diodes are irreparably damaged once that happens, but there are diodes that are actually designed to operate in the breakdown region of the diode characteristic curve. These are called zener diodes and they are discussed in more detail later on in this chapter. Ordinarily, operation in this region is avoided since the diode can no longer prevent current flow in the reverse direction when this takes place. The last part of the figure is the horizontal line to the right of the vertical line. This is the region where a PN junction diode is normally operated. This line represents forward voltage or V_F. In the case of the silicon diode, current begins to flow when the barrier voltage has been overcome by the external voltage source, or at about 0.7 volts. At this point, the forward bias voltage begins to equal the barrier voltage of the diode. Notice that the forward current begins to increase rather rapidly in an exponential manner. Once the barrier voltage has been overcome by V_F, the diode begins to conduct current in the forward direction, acting very much like a closed switch. Prior to reaching this point, the diode acts like an open switch, keeping current from flowing.

Now take a look at the characteristic curve of the germanium PN junction diode, shown in Fig. 2-11. Notice that this characteristic curve is very similar to that of the silicon PN junction diode. The apparent difference between germanium and silicon diodes can be

27

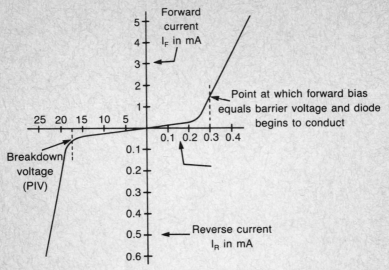

Fig. 2-11. Typical V-I characteristic curve of a Germanium PN junction diode.

seen easily in Fig. 2-11 by looking at the forward bias line. Here the germanium diode begins to conduct current in the forward direction when the barrier voltage of approximately 0.3 volts has been overcome. Of course, diodes vary from one to another, so 0.3 volts is a typical voltage that is specified for germanium diodes. The important aspect to note here is that this voltage is much lower than that of the silicon diode. Notice too the rapid increase in forward current once the barrier voltage has been surpassed by forward biasing the diode above 0.3 volts. In fact, between 0.3 volts and 0.4 volts, the forward current increases from 1 mA to approximately 5 mA. This is quite a jump considering the forward current remained below 1 mA up until the forward bias voltage reached 0.3 volts. However, once the diode's breakdown voltage is reached, forward current conducts rapidly in a very linear fashion. This is one reason for a series limiting resistor.

Another interesting aspect to examine with the germanium diode characteristic curve is the reverse bias breakdown voltage. In the silicon diode this is approximately 45 volts. In the germanium diode, this voltage, V_R, is approximately 20 volts. As in the silicon diode, operating in this region can damage the germanium diode. In practice, manufacturers specify the maximum safe reverse voltage that can be applied to a diode. This is called the diode's peak inverse voltage, or PIV. Also specified is the maximum allowable I_F for safe operation to prevent the diode from burning up because of too much current. Diodes also have a wattage rating similar to resistor wattage ratings. This is especially true in the case of zener diodes.

TEMPERATURE EFFECT

Temperature plays an important role in diode performance. In fact, later on you will see that this same effect has a great influence on the design of other solid-state circuits using both diodes and transistors.

The best way to show the effect of temperature on diode operation can be seen in the forward bias voltage region. Here, as temperature increases, forward breakdown voltage of the diode decreases. This means that less voltage is needed in the forward biasing mode to cause the diode to conduct current in the forward direction. As temperature increases, less V_F is needed to maintain the same forward current or I_F. This is termed negative temperature coefficient and is a destructive element in diode operation. As temperature within the diode begins to increase due to its own internal current flow, the forward current begins to increase, heating the diode even more and causing more forward flow. Finally, the forward current of the diode is exceeded and the diode is destroyed. This series of fatal events is termed thermal runaway and is seriously considered in the design of solid-state circuitry particularly with transistor amplifiers.

Another result of increased temperature can be seen by looking at the reverse current, I_R, curve of Fig. 2-11. Typically, for every 10 degree increase in temperature the reverse current doubles. This holds true for both germanium and silicon diodes, and is also a factor in circuit design when considering the effects of leakage current on circuit operation.

SYMBOLS

Thus far the diode has been represented by using a simplified drawing of the physical construction of the PN junction. However, it is simpler and more appropriate to represent the diode, as well as all other solid-state devices, using a symbol. The device is then easily recognizable in a schematic diagram.

Figure 2-12 shows the symbol of a diode and compares it to the symbol that was used previously. Notice that the direction of forward current is opposite the direction in which the arrow is pointing. Figure 2-13 shows how the symbol is placed in a circuit

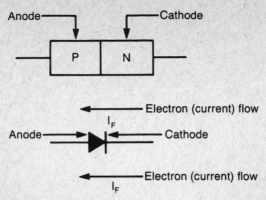

Fig. 2-12. Comparison of a PN junction diode and its symbol with current flow opposite to the arrow in the symbol.

allowing it to be forward biased. Here, the current flows from the negative side of the voltage source to the cathode of the diode, through the diode to the anode, and finally back to the positive side of the voltage source, in this case, a battery. Figure 2-14 represents a diode that is being reverse biased. In this figure either the diode symbol or the battery symbol could have been reversed. In each case, a current limiting resistor is connected in series with the diode. In the case of the reverse biased diode, only a very small leakage current flows represented by the dashed lines.

DIODE TESTING

There are actually two types of tests that provide you with information on the performance of a diode. The first is a functional test which tells if the diode is functioning properly; that is, wheth-

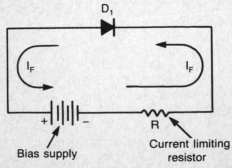

Fig. 2-13. A forward biased PN junction diode circuit using the diode symbol.

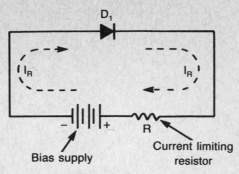

Fig. 2-14. A reverse biased PN junction diode circuit using the diode symbol with an extremely small flow of leakage current.

er the diode is shorted or open. It tests the quality of the diode and gives a quick analysis as to whether a diode is simply good or bad. To determine if the diode is performing according to the manufacturer's specifications, another test is performed called a quantitative test. It involves checking such specifications as V_F and I_R. Examine both of these testing procedures more closely.

Functional Testing

The functional type of testing that is performed on a diode is the one that is probably used the most in testing diodes, because in most cases you are concerned with only whether or not a diode is good and if it will work in a circuit. Chances are that you have already looked up the specifications of a particular diode you have in hand and wish to use it in a circuit, but you want to make sure it will function properly. Also, you may have removed a suspect diode from a circuit and want to test it to make sure that it is not the cause of the problem you may be having with that particular circuit.

As you have seen, when a diode is forward biased it conducts current and when it is reverse biased, it blocks current flow. In those examples of biasing, an external voltage source such as a battery, was used. The same thing can be done by using an ohmmeter. An ohmmeter provides the voltage source needed to forward bias a diode and shows on a meter scale or a digital display whether the diode is conducting or not. It therefore provides a voltage source and a means of viewing the good/bad status of the diode. Figure 2-15 is an illustration of how to forward bias a diode using an ohmmeter.

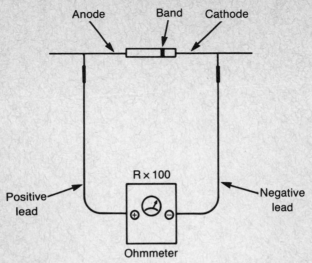

Fig. 2-15. A diode can be forward biased by the use of an ohmmeter which is also used to check the condition of the diode.

The first step is to make sure that the black lead of the ohmmeter is the negative lead and that the red lead of the ohmmeter is the positive lead. When using a digital meter this is usually true, but some ohmmeters using a meter movement indicator have the internal battery reversed; in other words, the black lead connected to the common terminal of the meter is actually the positive side of the internal battery of the meter and the red lead connected to the positive terminal of the meter is actually the negative side of the internal battery. This may sound confusing at first, but remember that ohmmeters are really designed to measure the resistance of a resistor or other resistive device and therefore do not take polarity into consideration. However, some digital multimeters have a special button or switch setting specifically for diodes.

As shown in Fig. 2-15, when testing a diode for forward resistance, the black/negative lead is connected to the cathode and the red/positive lead is connected to the anode. The meter range switch is set to R × 100. If the diode is not open or shorted, it usually reads a few hundred ohms. If it is open, the reading is the same as if the two meter leads were not connected to the diode. If it is shorted, the meter shows almost no resistance, or a minimal amount at best.

Figure 2-16 shows how to check a diode's reverse resistance to see if the diode is good or bad. In this test, the leads of the meter are reversed. The black/negative lead of the meter is connected to the anode, and the red/positive lead of the meter is

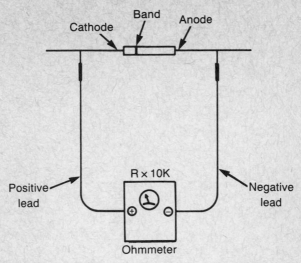

Fig. 2-16. In this test setup, the diode has been reversed.

connected to the cathode. The meter range switch should be set to R × 10 k. An indication of approximately 5 Mohms or more is typical of the standard silicon PN junction diode. In this test you will reverse bias the diode but not enough to break down the diode junction. There simply is not enough voltage in the battery to cause the diode to break down in the reverse direction.

Quantitative Testing

Quantitative testing is usually performed by the manufacturer prior to shipping diodes to a customer because certain test conditions are required that normally are not found in the typical workplace. This may require that temperature be precisely controlled since V_F and I_R are dependent upon temperature and vary as temperature varies. Most circuits are designed around these two parameters. Also, a current supply and a precision ammeter that reads down into the microampere range is needed, to test forward breakdown voltage and read leakage current, respectively.

Figure 2-17 shows how a diode is configured to be tested for I_R. Here, a specific reverse voltage is applied to a diode and the leakage current through the diode is measured.

Figure 2-18 shows how a diode is configured to be tested for V_F. In this test a specific level of current is forced through the diode in the forward direction while the voltage across the diode is measured.

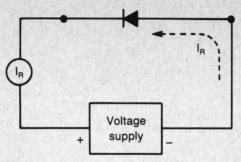

Fig. 2-17. One type of quantitative test is the measurement of reverse current through a diode with a specified reverse voltage on the diode.

APPLICATIONS

One of the most popular uses of PN junction diodes is in the rectification of alternating current. Rectification is the process of converting alternating current into direct current. A diode used to perform this function is appropriately called a rectifier. Most electronic circuits use dc voltage in their operation and therefore the ac voltage must first be converted to dc before it can be used. The diode usually operates with other devices, such as a transformer when used to supply power and operate as a dc voltage source. Figure 2-19 is an illustration of a typical half-wave rectifier circuit.

Notice that this circuit requires the addition of a transformer to either step up or step down the input voltage to that which is needed. This input voltage at the secondary of the transformer is then applied to the diode and resistor. The resistor is actually a

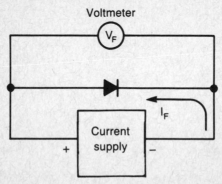

Fig. 2-18. Testing a diode that is functionally operational in the forward direction to measure the forward voltage drop is another type of quantitative measurement.

34

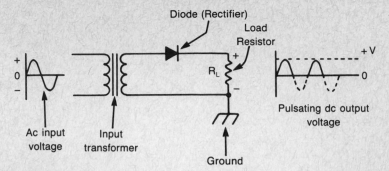

Fig. 2-19. A half-wave rectifier circuit can be as simple as this, or it can contain additional components to change the output dc pulses to more closely resemble a steady dc voltage output.

representation of the overall circuit resistance and is called a load. The diode only conducts when an alteration of the input voltage causes the diode to be forward biased. This occurs on every other alternation of the input voltage and gives an output of pulses as shown. This is called pulsating dc and is not much use in electronic circuits, but it is, however, a beginning to obtaining dc. The pulses in this circuit are positive with respect to ground because of the direction of current flow through the load resistor. The pulsating dc voltage could just as well be negative simply by reversing the diode.

To obtain a smoother, more continuous flow of dc voltage, other components such as a capacitor and another diode can be used with the circuit of Fig. 2-19, but these variations to the half-wave rectifier are discussed further in Chapter 11.

Some other applications of diodes are in their use as switches in digital switching circuits, detectors in FM radio circuits, and as protection devices in electronic circuits helping to prevent damage to other more expensive critical devices.

Chapter 3

Special-Purpose Diodes

There are a number of different types of diodes other than the PN junction diode. These other diodes are similar in construction to the typical diode, but because of their slight differences, are able to be used where the standard PN junction simply does not work. This discussion of special purpose diodes begins with the zener diode and includes a device called the Schottky diode which has expanded tremendously the applications of the integrated circuit or IC. Chapter 5 focuses on a special diode called an LED (light emitting diode) that emits photons of light as it conducts current.

ZENER DIODES

You have learned that an ordinary PN junction diode does not normally operate in the reverse voltage breakdown region because it may be damaged due to a high reverse current once breakdown has occurred. This high reverse current produces excessive heat and can very quickly destroy a diode. There are however, special diodes that do operate at or above the reverse breakdown voltage of the diode. These are called zener diodes. This is an important type of diode because of its ability to maintain a constant dc voltage drop across it, making it very useful in power supply applications.

The easiest way to view the operation of the zener diode is by using the standard characteristic, or V-I, curve shown in Fig. 3-1.

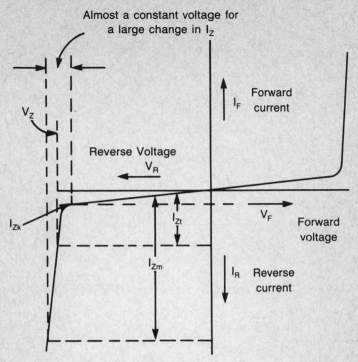

Fig. 3-1. A typical zener diode V-I curve showing a very small change in V_R for a very large change in I_R.

Notice the similarity between this curve and that of the typical PN junction diode. The difference is in the operating region of the zener diode, which is in the reverse voltage breakdown region. In contrast, the PN junction diode is constructed to operate in the forward voltage breakdown region. With the zener diode, as the reverse voltage is reached, the reverse current begins to increase slightly. Up until this point the reverse current has increased very little. Once breakdown occurs, reverse current increases sharply as shown. The curve shows that for a very small change in V_R, there is a very large change in I_R because the internal resistance of the diode drops to a negligible value when breakdown occurs. The diode is now said to be operating in its zener region and the reverse current at this operating point is called avalanche current.

Like the PN junction diode, the zener diode is very quickly destroyed if excessive current is allowed to flow. Therefore, a current limiting resistor is used to allow only a specific reverse current to flow, called a zener test current, or I_{Zt}. The I_{Zt} is always less than the maximum reverse current that the zener can safely support. All reverse current is labeled I_Z as shown in Fig. 3-1. The maximum allowable current is specified as I_{Zm} and the point at which breakdown current occurs is written as I_{Zk}, the zener knee current.

It is also important to note that zener diodes can be constructed so that they operate at various reverse voltages. Some of the more common voltages are 4.7 volts and 9.2 volts. Voltages may even go as high as several hundred volts. This is the operating voltage, V_Z, of the zener diode at I_{Zt} and not the reverse breakdown voltage itself.

Since the zener diode requires very little change in reverse voltage to effect a very large change in reverse current, it makes an excellent voltage regulator.

Figure 3-2 is a simplified version of a zener voltage regulator. The series resistor is the important aspect of this circuit because it limits the amount of current flowing through the zener. It is selected to allow the right amount of current to flow through the diode so that it operates in the zener breakdown voltage region. The zener diode is connected so that it is reverse biased. The input voltage must be higher than the reverse breakdown voltage of the zener or breakdown will not occur. Since the resistor and zener are connected in series, the sum of the voltage drops across those two components must be equal to the applied voltage. This is simply a restatement of one of Kirchhoff's voltage laws. The zener diode drops a voltage across it equal to its rated V_Z and the resistor drops the rest of the applied unregulated dc voltage. Because the zener diode is operating within its zener region, a wide range of input voltages and thus currents do not affect the voltage drops across the zener diode. As the input current varies, the voltage drop

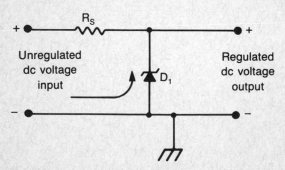

Fig. 3-2. A simplified version of a zener diode voltage regulator.

across the zener diode remains relatively constant. But the varying current is also flowing through the resistor. It is the resistor that varies its voltage drop as the input voltage varies, so the voltage across the zener diode is regulated or relatively constant. This voltage can be used as a voltage source for an electronic circuit. There is however, one problem, and that is the fact that electronic circuits sometimes require varying current for operation. An audio amplifier, as an example, has continually varying currents and therefore requires a power supply that can provide a constant voltage under these varying currents. Remember, the zener voltage regulator provides the constant voltage needed under varying current conditions. However, there is a minimum and maximum, or current range, that must be maintained.

The circuit of Fig. 3-3 gives you an example of a simplified zener voltage regulator circuit. It is similar to that of Fig. 3-2 except that a load resistor, R_L, has been added. This load represents the electronic circuit that would be connected to the output of the regulator. The total current flowing through the regulator circuit is the combination of current flowing through the zener diode and the current flowing through the load, or $I_Z + I_L$, respectively. This is the current that flows through the series limiting resistor R_S. R_S must be just the right size to keep the zener diode within its operating region and yet still allow the proper amount of I_L to flow through the load.

Ordinarily, if the zener was not in the circuit and R_L changed in value, as does happen in circuits, the output voltage would also change, or vary, according to changes in R_L. Adding the zener prevents the output voltage from changing as R_L changes. As an example, if the current through R_L decreases, the voltage across

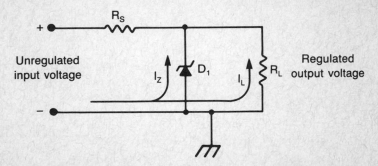

Fig. 3-3. In this circuit where $I_T = I_Z + I_L$, R_L is a load resistance equivalent to an electronic circuit receiving dc voltage from this zener regulator.

R_L also decreases. But the zener is in parallel across R_L and its voltage remains relatively constant with any changes across R_L. However, as I_L decreases, I_Z tends to increase, keeping the total current through R_S the same as before. This means that $I_Z + I_L$ remains constant with changes in R_L.

Designing a zener voltage regulator is rather easy. There are a few parameters that must be known, put into an equation, and then solved. As an example, assume you would like a voltage regulator with the following parameters:

Input voltage: 15 volts - 18 volts
Output voltage: 9.1 volts
Output load current: 0 - 50 mA
First calculate R_S.

$$R_S = \frac{V_{in(min)} - V_Z}{1.1\ I_{L(max)}}$$

$$R_S = \frac{15 - 9.1}{1.1(.05)}$$

$$R_S = 107\ ohms$$

Since 107 ohms is not a standard value resistance, the value of 100 ohms is chosen. The next lower value is selected so that I_Z will not drop below the value necessary to keep the zener diode within its breakdown region.

Secondly, the maximum power that can safely be dissipated by the zener diode must be found. This is $P_{Z(max)}$ and is found as follows:

$$P_{Z(max)} = V_Z \frac{V_{in(max)} - V_Z}{R_S} - I_{L(min)}$$

$$P_{Z(max)} = 9.1 \frac{18 - 9.1}{100} - 0$$

$$P_{Z(max)} = 0.81\ watts$$

Since 0.81 watts is not a standard wattage rating for a zener diode, the next highest rating is used. In this case, a 1 watt zener diode would be used.

TUNNEL DIODES

The characteristic curve of the tunnel diode is shown in Fig. 3-4. This diode exhibits a V-I curve that is considerably different from the characteristic curves of the ordinary PN junction diode and the zener diode.

As you can see, there is a very large reverse current for a very small reverse bias voltage. Similarly, a very small forward bias voltage produces a large forward current flow almost immediately. This is due mainly to the heavily doped PN junction. Although this causes an extremely high internal barrier voltage as compared to the typical PN junction diode, once a small forward bias voltage is applied, electrons are forced through the tunnel diode's very narrow depletion region of a high velocity due to the large ion charges on each side of the junction. It's almost as though the electrons, because of their high velocity, are tunneling their way through the barrier voltage from one side of the junction to the other. (Thus the term tunnel diode.)

There is a point reached where an increase in forward bias voltage does not increase forward current. In fact, forward current

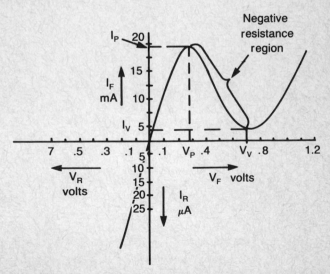

Fig. 3-4. The V-I curve for a tunnel diode showing the negative resistance region.

begins to decrease. As shown in Fig. 3-4, forward current reaches a peak, I_P, then dips into a valley as forward voltage increases. Its minimum current at this point is referred to as valley current, or I_V. Once again as forward voltage continues to increase, forward current increases, only this time, much the same way forward current increases in an ordinary PN junction diode with an increase in forward voltage.

The time during which the forward current decreases and forward voltage increases is referred to as the negative resistance region because it actually contradicts Ohm's Law which states that as voltage increases, current increases, providing resistance remains constant. However, in this case, in this region, current decreases as voltage increases, as though resistance had a negative effect on the current. Tunnel diode action occurs during this period of time.

Construction

Tunnel diodes have been constructed using silicon and germanium as are most ordinary PN junction diodes. However, today most tunnel diodes are manufactured from gallium arsenide, GaAs, or gallium antimonide. A popular method of forming the PN junction within the diode is by the alloyed method. This is a widely used method in forming PN junctions. Stated simply, it is a process whereby a semiconductor material is placed on another semiconductor material. The first material is melted and dissolves a small portion of the material it is sitting on. It is then said to alloy with the material and form a junction.

On the outside, tunnel diodes appear to be very similar to any other PN junction diode. However, they usually are quite small in size, usually about the size of a match head for the lower current types.

The symbols used for the tunnel diode are shown in Fig. 3-5.

Applications

The faster a diode begins to conduct forward current with forward voltage applied, the faster the diode turns on or changes from the off-state to the on-state. In the case of the tunnel diode, this makes it extremely attractive in oscillator and computer applications where high speed switching is absolutely essential. When used as an oscillator, it is capable of operating in the gigahertz region. As a switch, it can change states in only a few nanoseconds.

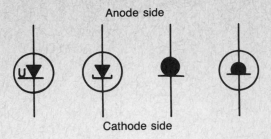

Anode side

Cathode side

Fig. 3-5. Some typical symbols used to represent the tunnel diode.

Figure 3-6 is an illustration of the use of the tunnel diode in an oscillator circuit.

Here, the tunnel diode is used to help maintain the oscillations of the resonant tank circuit, L_1 and C_1. R_1 and R_2 ensure that the diode's operating voltage and current remain in the negative resistance region. With power applied, oscillations are produced in the LC circuit. These oscillations cause a voltage across the LC circuit which in turn produces a shift in the tunnel diodes operating point. The diode's resistance then changes allowing current through the diode to sustain the circulating current of L_1 and C_1, producing a high frequency alternating current signal output. It is the negative resistance region which maintains these oscillations within the LC circuit.

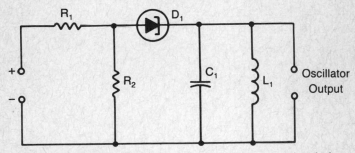

Fig. 3-6. An oscillator circuit using a tunnel diode as the active device.

Varactor Diodes

The varactor diode is actually used more as a variable capacitor than as a diode. In the PN junction diode, the depletion region acts to separate the P and N sides of the diode. In this respect it separates or insulates the two sides of the diode in much the same

43

way that a dielectric separates the two sides of a capacitor. However, with a diode, the depletion region varies as the bias voltage across the diode varies. Thus, the capacitance of the depletion region also varies. In the case of ordinary PN junctions, the internal capacitance of the depletion region is very small, and not really suitable for capacitive action. However, certain diodes, called varactor diodes, are constructed making use of this principle.

Figure 3-7 shows the capacitance voltage characteristic curve of the typical varactor diode.

This is not a typical V-I characteristic curve since capacitance and voltage are being compared rather than currents and voltages. Therefore no particular part of the V-I curve is being viewed, only a relationship between two parameters.

As the reverse bias is increased, the diode capacitance decreases. This is because as V_R is increased, the depletion region within the diode becomes larger. This is analogous to the thickness of the dielectric in a capacitor. As the thickness increases, the capacitance decreases. In the varactor diode, this thickness is represented by the depletion region. Since it increases as V_R increases, then capacitance decreases as in a standard capacitor. Stated another way, the varactor diode's capacitance varies inversely to the applied reverse bias voltage. It does not vary in a linear manner and therefore is not inversely proportional to the applied bias voltage. In fact, if the diode is forward biased, its capacitance continues to increase, since the depletion region now gets

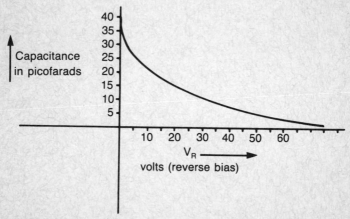

Fig. 3-7. The characteristic curve of a varactor diode is not the same as other V-I curves because the device acts like a capacitor rather than like an ordinary PN junction diode.

smaller in size, until the internal barrier voltage is overcome and the diode begins to conduct in the forward direction. Remember too that the diode does not operate effectively as a varactor once its reverse bias breakdown voltage is reached.

The capacitance of the PN junction itself is found from the following equation:

$$C = \frac{\epsilon A}{d}$$

C = Capacitance of PN junction
ϵ = Dielectric constant
A = Area of P and N material
d = Width of depletion region

Most varactor diodes have a very high Q. As with any capacitor, as the frequency, capacitance, or series resistance decreases, the Q of the device increases. The series resistance is actually the internal resistance of the device as seen by an ac signal passing through the diode. This relationship is expressed as follows:

$$Q = \frac{1}{2\, fCR_S}$$

The higher the Q, the better the efficiency of the device, whether a capacitor or a diode.

What this equation also means is that the Q of the diode is proportional to the reverse bias of the applied voltage. As V_R increases, Q increases. Since the device operates as a variable capacitor up until the forward breakdown voltage is reached, the Q is at its highest just before this point. There is also a frequency operating point at which the Q of the diode can be reduced to 1. This is the cutoff frequency, expressed as follows:

$$f_{CO} = \frac{1}{2\, CR_S}$$

Since the varactor diode operates in the reverse bias voltage region, leakage current can also have an effect on the Q of the diode.

45

The maximum leakage current I_R is usually specified by the manufacturer. This is because Q is inversely proportional to I_R. As I_R decreases, the Q increases. Leakage current is usually low in silicon devices and even lower in gallium arsenide devices and therefore the latter material is more widely used in the construction of varactor diodes.

Construction

The type of construction of the varactor diode is usually a function of the application for which it is intended. High frequency, microwave varactor diodes are very small in size and are constructed to take stray capacitance and inductance into consideration. This is because at microwave frequencies even the length of a lead of wire affects the proper operation of an electronic circuit.

As with ordinary PN junction diodes, varactor diodes also have power ratings. The power ratings of varactor diodes are not dependent upon dc as in the PN junction diode but upon ac instead. That is because the reverse biased varactor diode passes ac while blocking dc. Therefore the power that a varactor diode dissipates is a function of the ac current passing through the series resistance, R_S, of the diode.

In addition, there are several symbols that are used to identify the varactor diode, shown in Fig. 3-8.

Fig. 3-8. Some of the symbols used to represent the varactor in electronic circuits.

Applications

Probably the most popular application of the varactor diode is in the control of frequency in electronic circuits. Figure 3-9 is a circuit in which a varactor diode is used for just this purpose.

The varactor diode is reverse biased. That is because in this circuit the frequency of the resonant circuit, D_1 and L_1, is deter-

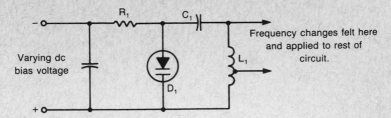

Fig. 3-9. The varactor diode in this circuit is used to control the frequency of resonant circuit D_1L_1.

mined by the amount of reverse bias on D_1. C_1 is relatively large and therefore is used only to block dc from flowing through the coil. As the dc bias voltage is varied, the capacitance of the varactor diode varies, thus varying the frequency of the resonant circuit.

In this application, varactor diodes can be used in AFC (automatic frequency control) circuits in FM radio. Used in this way, a feedback signal consisting of a dc voltage is taken from the output of the FM circuitry. If the output frequency varies above or below the required frequency, this change or error in frequency is sensed as a dc voltage. It is then fed back to the varactor diode and is used to change the original frequency in the tank circuit by varying the reverse bias on the diode. Figure 3-10 is a simplified block diagram showing this principle of operation.

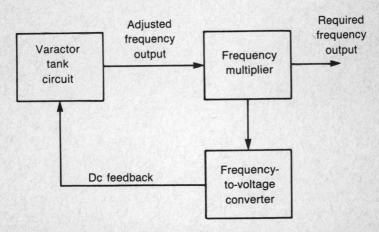

Fig. 3-10. A simplified block diagram showing how frequency is changed to voltage then fed back to change the bias on the varactor diode and thus, the frequency output.

THE SCHOTTKY DIODE

The Schottky, or hot carrier, diode received its name from the German scientist Schottky in 1938. It was he who discovered its operating principle based on the special effect of the barrier voltage within the diode. This diode is therefore sometimes referred to as the Schottky-barrier diode.

In the ordinary PN junction diode, the barrier voltage is approximately 0.7 volts. In the Schottky diode this barrier voltage is about 0.3 volts. Also, the leakage current, I_R, is very small in this diode since all action within the diode takes place with majority carriers rather than with minority carriers.

The operation of this diode is similar to an ordinary PN junction diode except that the movement of electrons across the junction from cathode to anode is very similar to the movement of electrons in a vacuum tube. In the vacuum tube the cathode is heated and emits electrons that possess a high level of kinetic energy. The electrons passing through the Schottky diode possess the same kind of energy and are similar to the hot carriers just mentioned, hence the name hot carrier diode.

Construction

The junction of the Schottky diode is formed by joining an N type semiconductor with a metallic material such as gold or silver. This is the reason for the difference in barrier voltage between this diode and the ordinary PN junction diode. Figure 3-11 shows a simplified illustration of the HCD (hot carrier diode). Also shown is the symbol for an HCD. Do not confuse this with the symbol for a zener diode.

Applications

The Schottky diode has found many applications in high frequency electronic circuits because of its ability to change states, from off to on and back again, at a much faster rate than the ordinary PN junction diode. In Chapter 13 you will see that Schottky barrier diodes, sometimes referred to as SBDs, are use to increase the switching speed of transistor-to-transistor logic integrated circuits.

HIGH FREQUENCY DIODES

Although the past few diodes that have been discussed are used in high frequency applications, there are other devices constructed

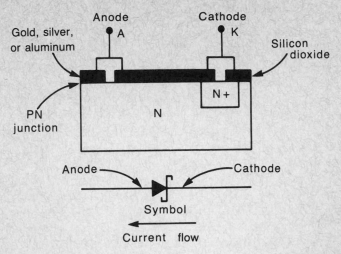

Fig. 3-11. A simplified illustration showing the basic physical construction of the Schottky diode and its symbol.

specifically for high frequency use. These are the PIN diode, IMPATT diode, and the Gunn-Effect diode.

The term PIN is an abbreviation for P type semiconductor, Intrinsic semiconductor, and N type semiconductor and is constructed as the name implies. This is shown in Fig. 3-12.

The intrinsic (pure or non-doped) semiconductor material is placed between two other semiconductor materials. The advantage of this type of diode is its ability to change from a nonconducting state to a conducting state at a very rapid rate or frequency. This makes the PIN diode suitable for high frequency applications in the microwave range.

When used as a high speed switch the diode is turned on and off by subjecting it to forward and reverse biases, respectively.

Fig. 3-12. The PIN diode's construction concept is shown here.

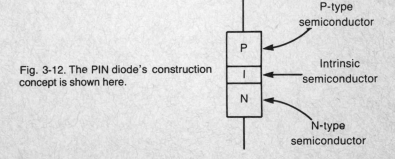

49

Because of its wide internal resistance range, which is very linear with respect to forward bias, it is easily controllable by low currents and bias voltages. This allows the PIN diode to be used as a current controlled resistor meaning that the current flowing through it will determine the resistance of the device. As an example, a current change from 1 microampere to 100 milliamperes causes an internal resistance change of the device of from 10 k ohms to 1 ohm when forward biased lending it as an attenuation device in audio limiter circuits.

IMPATT is an abbreviation for Impact Avalanche Transition Time. This means that this particular device has a very sharp breakdown knee in the reverse bias region of its operating curve and makes an excellent rf power generator or oscillator. A typical IMPATT diode can operate at several watts and at a frequency as high as 10 gigahertz. Although the efficiency of the device is only about 10 percent, it is still widely used. Today, newer devices are being developed that can be operated with greater power and efficiency.

IMPATT diodes are usually very small in size. This is because the metal cavity in which the diode is mounted becomes part of the resonant circuit itself at microwave frequencies, and the case serves as a heat sink to conduct harmful heat away from the very small PN junction inside of the diode.

The Gunn-Effect diode is not truly a PN junction diode as is true with many high frequency devices. It is similar to the IMPATT diode as far as packaging and use, and it operates in its negative resistance region giving it the property of producing extremely high frequencies when properly reverse biased.

The Gunn-Effect diode is typically manufactured using N-type gallium arsenide semiconductor crystals and does not contain a PN junction like the ordinary PN junction diode. The device can be damaged if it's not biased in the direction for which it is specified.

THE SHOCKLEY DIODE

The Shockley diode is a four-layer PNPN diode containing three junctions and is named after its inventor William Shockley. The characteristic curve of the device is shown in Fig. 3-13.

Notice that a point in forward voltage is reached where a breakdown occurs in the negative direction. This is referred to as the minimum switch-on voltage, V_S, or tripping potential. This happens because as forward voltage begins to increase, junction

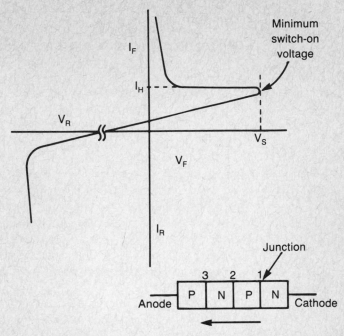

Fig. 3-13. The characteristic curve for the Schottky diode along with a simplified drawing of the construction of this PNPN junction diode.

two remains reverse biased allowing only minority current carriers to flow. Once this junction breaks over the diode switches on and current flows in the forward direction. Notice the negative resistance characteristic that the diode displays. This shows that current increases slightly as voltage decreases and therefore the diode can switch rapidly from the on or the off state. I_H is the minimum hold-on current needed to keep the device conducting. Once the diode is switched on only a very small voltage remains across the diode and is typically the sum of the knee voltages of two junctions in series, or about 1.4 volts dc.

VARISTORS

The varistor is a diode which displays a nonlinear resistance when voltage across it is increased or decreased. Varistors are used as protection devices in electronic circuits where a large voltage spike might be expected while the circuit is operating.

Typically, when voltage across an ordinary resistor increases, the current through that resistor increases proportionately with the voltage, since resistance remains constant. However, with a varistor, current increases by perhaps a factor of ten, depending on the manufacturing properties of the varistor. This means that the resistance of the device decreases tremendously with only a slight increase in voltage. Varistors can be used as lightning arrestors. The high voltage surge of lightning causes the resistance of the varistor to decrease, allowing the lightning to pass through it rather than the equipment. Once the surge has passed, the resistance of the varistor returns to normal. They are especially useful across motors and other inductive windings where a voltage kick is experienced when the circuit is opened.

General Electric makes a varistor called a Thyrector made from the semiconductor material selenium. The Thyrector works like a zener diode which acts as an open until a specific reverse breakdown voltage is reached. Once that happens, the diode acts like a closed switch, routing voltage away from the rest of the circuit. This type of varistor is usually connected in series in the opposite direction from a second device to catch voltage surges in either direction.

As with most diodes, the varistor can be checked using an ohmmeter. However, most varistors are constructed so that they appear to be open in both directions. This is because they are normally manufactured bidirectionally in one package. It is not uncommon to see a varistor that is badly burned on a printed circuit board because of lightning damage. Typically, a varistor may be rated for 115 volts or 130 volts or more, but lightning surges on a 115 volt ac line almost assuredly damages a varistor. In that case

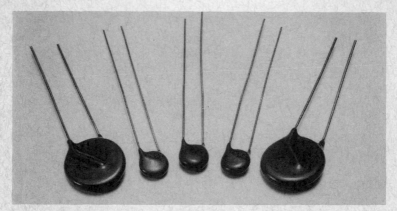

Fig. 3-14. Varistors come in many sizes and shapes as shown by these manufactured by G.E. (John Sedor Photography).

it is usually expedient to replace all varistors on the printed circuit board or other piece of electronics equipment that has experienced a lightning surge.

Figure 3-14 is a photo of some varistors made by GE. Notice the two different sizes. Varistors are available in many different sizes depending primarily on their surge voltage rating.

Chapter 4

The Bipolar Transistor

A bipolar transistor is a solid-state device containing three semiconductor materials that is used to control current flow in an electronic circuit. This is achieved by varying the amount of voltage on each of the three regions of the device. Bipolar transistors are very similar to the PN junction diode, and are usually termed junction transistors or simply transistors.

The transistor is constructed from germanium or silicon just as the PN junction diode is, but another layer is added, giving two junctions instead of one. There are two methods in which these three semiconductor materials may be configured. The first is by placing a P-type material between two N-type materials and the second is by placing the N-type material between the two P-type materials.

NPN AND PNP CONSTRUCTION

In the configuration of Fig. 4-1, the P-type material is sandwiched between the two N-type sections and is called the base region or just the base of the transistor, while the outer two sections are called the emitter and the collector. In practice the base is doped much more lightly in respect to the emitter and collector and is also considerably thinner. And, as with the ordinary PN junction diode, leads or wires are attached to each section so that an electrical connection can be made.

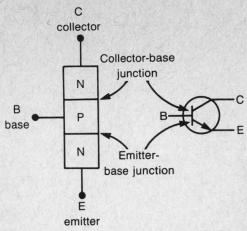

Fig. 4-1. An NPN transistor's construction and its symbol; the emitter is identified by the arrow.

In Fig. 4-1 the emitter, base, and collector are marked, but this may not always be the case. The emitter can always be recognized as the lead with the arrow.

The other method in which a transistor can be presented is shown in Fig. 4-2.

Here, the N-type semiconductor material is sandwiched between the two P-type semiconductor materials. This is known as a PNP transistor and, like the NPN transistor, the base region is

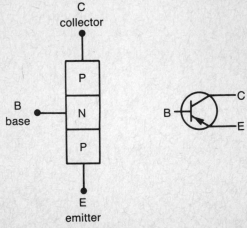

Fig. 4-2. A PNP transistor's construction and its symbol, this time with the arrow pointing inward.

very thin and lightly doped. Again, the leads are marked as emitter, base, and collector. Notice that the emitter of the PNP transistor is also identified with an arrow, but pointing in the direction opposite the arrow in the emitter lead of the NPN transistor.

Although both types of transistors have the same material for the collector and emitter regions, these regions are not interchangeable because the emitter region of each transistor is more heavily doped and is actually bigger physically.

BIPOLAR TRANSISTOR CONSTRUCTION

The construction of the bipolar transistor is very similar to that of the PN junction diode. Usually, variations of certain methods are used. Figure 4-3 is an illustration of an epitaxial method used to construct a typical NPN transistor.

In this type of manufacturing a lightly doped P-type layer is epitaxially grown onto the top of an N-type semiconductor crystal. The P-type layer is the base of the transistor while the N-type semiconductor represents the collector. Finally, the emitter is formed by diffusing another N-type semiconductor material into the P-type epitaxial layer. Diffusion takes place by placing a pellet of N-type material onto the P-type base and heating the assembly. The pellet melts and diffuses into the base. After the heat is removed the two materials recrystallize to form a PN junction of emitter and base.

Another method is by complete diffusion of all three semiconductor materials, shown in Fig. 4-4. Here, an N-type semiconductor crystal is subjected to a trivalent impurity at a very high temperature. The trivalent then diffuses into the crystal and forms a P region. Next, a pentavalent element is diffused into the P region to form an N region, producing an NPN transistor. PNP transistors can be formed using the same method.

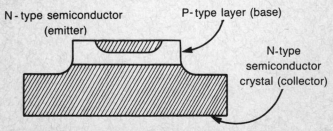

Fig. 4-3. In this method of transistor construction, the base is epitaxially grown onto the collector and then the emitter is epitaxially grown onto the base.

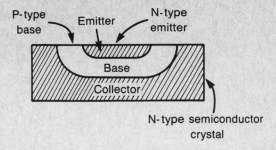

Fig. 4-4. Diffusion is a process of transistor construction used in manufacturing both NPN and PNP transistors with an NPN transistor shown here.

Finally, transistors may be constructed by the alloy method, shown in Fig. 4-5. This is a very simple process in which a pellet of pentavalent material is placed on either side of a P-type semiconductor material that serves as a base. The entire assembly is heated and the pentavalent diffuses into the P-type material. When the heat is removed from the assembly, the P-type material and the pentavalent recrystallize to form an NPN transistor.

If you'll remember, these construction methods are the same as those used in the manufacture of the PN junction diode. The main difference is in the physical size of the base as compared to the other regions and in the amount of doping in each region. The base is the smallest or thinnest of the three elements and is more lightly doped than either the emitter or the collector. The emitter however, is more heavily doped than either the base or the collector.

Also, transistors are not constructed individually. As with

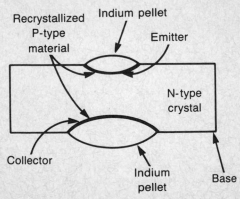

Fig. 4-5. This method of bipolar transistor construction is termed the alloy method with a PNP transistor shown here.

diodes, hundreds are constructed on round disks approximately two inches in diameter and then separated or cut usually by an industrial type laser, into individual transistor chips ready for testing and then packaging.

PACKAGING

A transistor, like any other electronic component, must be placed in a package suitable for use. The particular package is chosen for proper heat transfer away from the PN junctions to prevent damage to the transistor and for proper electrical connections and current handling abilities.

There are many package styles that are used in housing a transistor, as shown in Fig. 4-6. These different styles are normally referred to as outlines and are properly called transistor outlines or TOs.

Sometimes an extra lead is attached to the case along with the leads that are connected to the emitter, base, and collector regions of the transistor. Also, the package can be made from metal or from plastic. The larger type transistor packages are usually intended for those devices that must carry a large amount of current.

Most transistor packages are hermetically sealed, like the diode, to keep them free from dust and humidity, and the lead spacing

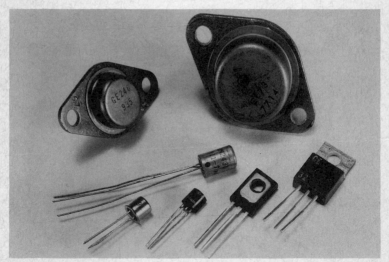

Fig. 4-6. Transistors come in a variety of packages, depending on power ratings and the circuit configuration in which they will be used (John Sedor Photography).

itself is usually standard from one manufacturer to another so that they are interchangeable in an electronic circuit. This means that a TO66 plastic package from one manufacturer can be replaced by a TO66 plastic package from another manufacturer on the same printed circuit board.

The larger packaged devices are usually mounted on the chassis which may house the electronic circuit in which it is used. This is so the amount of heat dissipated by the transistor heat sinks to the chassis and prevents the transistor from being damaged.

When encountering a transistor where emitter, base, and collector leads are not marked, it is best to consult a semiconductor cross-reference guide to determine which leads are the emitter, base, and collector for that particular TO style, and any other specifications needed to determine the applicability of the device in a given electronic circuit. A good transistor cross-reference guide gives lead identification for several different TOs, typical applications, current and voltage ratings, and also power consumption.

BIPOLAR TRANSISTOR ACTION

Bipolar transistors are used in a variety of ways, but in general they are used as electronic switches and in the amplification of voltage, current, or both. Doing so requires controlling the current through the transistor by varying the bias voltages connected to the emitter, base, and collector. Once these proper biasing conditions are met the transistor will function as intended.

NPN Biasing

The transistor consists of three semiconductor regions connected by two PN junctions. There is a PN junction between emitter and base forming an emitter-base junction. And there is a PN junction between the collector and base regions forming a collector-base junction. These are sometimes referred to as the emitter junction and collector junction, respectively, as shown in Fig. 4-7.

In a properly operating NPN bipolar transistor, the emitter junction is forward biased while the collector junction is reverse biased. The collector junction, like any reverse biased PN junction, supports a very small leakage current. These are the holes in the N-type collector and the electrons in the P-type base.

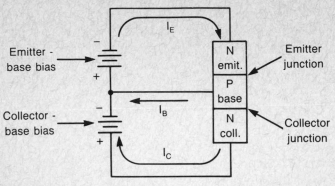

Fig. 4-7. Biasing for an NPN transistor is shown here where $I_E = I_C + I_B$.

Although the base is very thin and only lightly doped, the emitter junction still functions like a typical PN junction diode. Therefore the majority carriers of the base and of the emitter regions combine at the emitter junction. But, because of the much larger amount of majority carriers in the emitter, the base is not able to supply all of the holes needed to combine with the electrons from the emitter. However, most of the electrons flow from the emitter, through the base region, and into the collector region. Once here they are attracted to the large positive potential of the collector bias supply. Approximately 5 percent of the electrons from the emitter flow out of the base region while the remaining 95 percent flow through the collector region. This current is referred to as collector current and is designated I_C. The smaller amount of current flowing out of the base is referred to as base current and is designated as I_B. The current that flows from the emitter is called emitter current, designated I_E. Since I_E flows to form I_B and I_C, there is a direct relationship among the three parameters which can be expressed as follows:

$$I_E = I_B + I_C$$

and therefore:

$$I_C = I_E - I_B$$

Since the emitter junction behaves like an ordinary PN junction diode, there is a barrier voltage across this junction of from 0.3 volts to 0.7 volts dc depending upon the material of the semiconductors.

Also, the collector junction has very large reverse bias applied to it to effectively attract the emitter current. This means that the reverse bias applied to the collector junction is considerably higher in potential than the forward bias applied to the emitter junction. These are important aspects to keep in mind.

PNP Biasing

In biasing a PNP transistor, the same principles apply that were given for biasing an NPN transistor. The emitter junction is forward biased and the collector junction is reverse biased. Basically then, the external bias voltages are reversed for the PNP transistor and the majority charge carriers are holes rather than electrons. The biasing scheme for a PNP transistor is shown in Fig. 4-8.

TRANSISTOR AMPLIFICATION

Initially, the transistor was shown in a block form in discussing NPN and PNP biasing. However, in presenting the transistor as an amplifying device, the standard transistor symbol will be used. As shown in Fig. 4-9, the typical NPN transistor amplifier has a forward biased emitter junction and a reverse biased collector junction. The emitter bias voltage supply is normally labeled VEE while the collector bias voltage supply is identified as VCC. Also, the transistor is usually labeled as Q_1. These are the standard labels used throughout the electronics industry.

Notice in Fig. 4-9 that VEE has been made variable to show that by varying I_E, I_C also varies. Remember, $I_E = I_B + I_C$. If VEE is increased, more electrons are injected into the base region. This

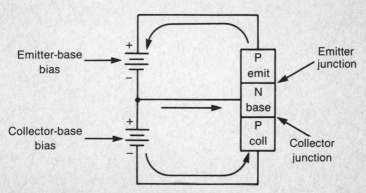

Fig. 4-8. As with the NPN transistor, this PNP transistor has its emitter junction forward biased and its collector junction reverse biased.

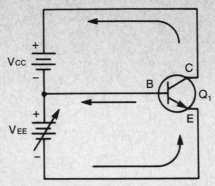

Fig. 4-9. A typical NPN transistor symbol in a properly biased circuit.

means I_E increases. If I_E increases then more electrons are going to flow through the collector, so I_C increases when I_E increases. Any electrons not flowing into the collector flow through and out of the base, and I_B also increases. So, simply by increasing I_E, I_C and I_B both increase. Now that you have seen how all of these currents are directly related to each other, it is necessary to take this concept one step further to demonstrate amplification.

Basically, amplification consists of a transistor producing a large output signal from a much smaller input signal by varying the current through the device. This idea is shown in Fig. 4-10.

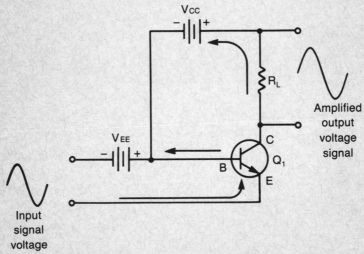

Fig. 4-10. A typical NPN transistor amplifier circuit.

The changes made in this circuit allow Q_1 to take a very small input voltage or current and produce a larger output voltage or current. Notice R_L in the collector circuit of Q_1. I_C flows through R_L producing a voltage drop. The input is connected between the emitter and VEE. A complete circuit is made here when the input signal is connected between the emitter of Q_1 and VEE.

R_L has a high value to drop a large voltage, but does not sufficiently limit I_C flowing through it because I_C remains almost as great as I_E.

To successfully amplify a signal, a small positive input signal is needed to add to the forward bias of VEE. This increases I_E and subsequently I_C. The voltage drop across R_L also increases. A negative input signal has the opposite effect and the voltage drop across R_L decreases. As you can see, the output voltage is following the input voltage, change for change, but the output voltage change is considerably greater in its swing from positive to negative because it is developed across a high value load resistor, whereas the input voltage is applied to the low resistance of the forward biased emitter junction. Also, VCC is much larger in value than VEE. In effect, a small change in input voltage causes a very large corresponding change in output voltage or, stated another way, an amplification of the input signal. In this case, a voltage has been increased from a very small value to a very large value resulting in voltage amplification. However, there are amplifiers designed to amplify currents as well as voltages. These are referred to as current amplifiers. Keep in mind that a PNP transistor amplifies in the same manner and could replace the NPN transistor shown in Fig. 4-10. Of course, the bias supplies must be reversed and the currents will flow in the opposite direction from that shown for the NPN transistor amplifier.

TRANSISTOR CIRCUIT CONFIGURATIONS

There are three basic circuit arrangements used in connecting a bipolar transistor for use as an amplifying device. Each has its advantages and disadvantages, which are summarized at the end of this chapter. The three circuit configurations are common base, common emitter, and common collector. The lead labeled as common is common to both the input and output signal paths. Usually the common lead is connected to circuit ground, thus giving the terms grounded base, grounded emitter, and grounded collector equal importance. But, no matter which circuit configuration is used, the emitter junction is still always forward biased while the collector junction is reverse biased.

Common Base Configuration

In the common base configuration, the base lead is connected to circuit ground while the emitter junction is used as the input and the collector junction is used as the output. This is basically the same circuit that was used to introduce transistor amplification and is shown again in Fig. 4-11.

This illustration shows an NPN and PNP transistor connected in a common base configuration. The only differences between these two circuits are the direction of current flow and the method for connecting the bias supplies. The base is common to both the input and output circuits. In this case the input signal is an

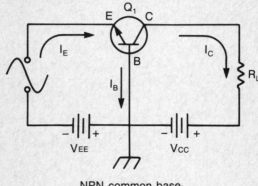

NPN common base

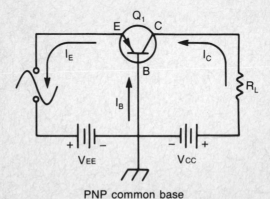

PNP common base

Fig. 4-11. The common base circuit configuration for both the NPN and PNP transistors.

alternating voltage and current. The load on the collector-base circuit can be resistive or an inductive load, such as a resistor or even a relay.

The advantages of the common base configuration are the fact that it makes a good voltage amplifier and also that the input signal in effect sees a very small input resistance. This is because the emitter-base junction is forward biased and a forward biased PN junction always offers a very low resistance to current flow. This is usually in the range of from 20 ohms to 50 ohms. In contrast, the collector junction is reverse biased and therefore has a much larger output resistance of approximately 100 k ohms to 1 megohm.

The common base amplifier offers voltage gain, but does not offer current gain. This is because collector current is always slightly less than emitter current. However, I_C is still large enough to offer a power gain, so the common base configuration offers low input resistance, high output resistance, voltage amplification, and power amplification.

Common Emitter Configuration

In the common emitter configuration it is the emitter that is connected to circuit ground, shown in Fig. 4-12. In this illustration the forward bias is now applied to the base instead of the emitter of the transistor. In this case V_{BB} controls I_B instead of V_{EE} controlling I_E. It is the base circuit now that is the input circuit and not the emitter circuit. The collector current however still flows

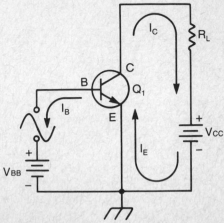

Fig. 4-12. A common emitter circuit configuration for an NPN transistor.

through the load, R_L. And the input signal is part of the base circuit.

It may not be apparent from Fig. 4-12 that the collector junction is reverse biased, but it is. This is because VBB causes the emitter-base junction to be forward biased offering a low resistance to current flow. Therefore VCC effectively sees only the collector-base junction. Since VCC is connected in a reverse biased manner, the collector-base junction is also reverse biased.

Also, the relationship among emitter, base, and collector currents still remains the same. That is, $I_E = I_C + I_B$. However, since base current is now the input current, this can be rewritten as $I_B = I_E - I_C$. In other words, a small change in base current effects a large change in collector current, just as in the common base configuration where a small change in emitter current caused a large change in collector current. However, since a small base current can control a large collector current, this amplifier can be used to amplify current as well as voltage. And, in turn, there is also the advantage of power amplification. In effect then, the common emitter configuration offers current, voltage, and power amplification. In addition, because the emitter-base junction is forward biased, it offers a relatively low input resistance, although not as low as the common base configuration. It is usually around 2 k ohms while the output resistance is approximately 70 k ohms.

Common Collector Configuration

In the common collector configuration it is the collector that is the common region between the input and output signals. The input signal of the common collector circuit appears across the collector base regions of the transistor. The output voltage is developed across the load resistor connected across the emitter and collector regions of the transistor, shown in Fig. 4-13.

If the input signal increases, I_B increases thereby increasing I_E. When this happens the voltage drop across the load increases by the same percentage as the input voltage. In fact, there is no voltage amplification because the voltage at the emitter is slightly less than the voltage at the input by an amount equal to the forward bias voltage of the emitter-base junction. So, although the voltage gain is slightly less than one, the emitter voltage tends to track or follow the input signal voltage. For this reason the common collector circuit is also referred to as an emitter follower.

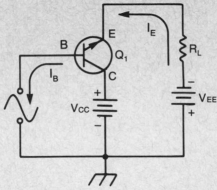

Fig. 4-13. A typical common collector circuit configuration using an NPN transistor.

There is however, current gain because I_E tracks I_B and is considerably larger. Therefore there also exists power gain in spite of the less than unity (less than one) voltage gain. Also in this particular circuit arrangement, the input resistance is relatively high because the input signal appears across a reverse biased collector-base junction and is usually around one megohm. The output resistance is much lower, around 500 ohms, because of the high emitter current flowing through the load.

The common emitter circuit configuration is usually used as an isolating device between transistors rather than as an amplifying device and is also referred to as a buffer. Its high input resistance and low output resistance make it ideal for this type of application.

CHARACTERISTIC CURVES

Just as diodes were viewed using characteristic curves, bipolar transistors can more easily be understood by looking at their characteristic curves. These are sometimes referred to as V-I (voltage-current) curves. There are V-I curves for each of the bipolar transistor configurations just presented. Each of these curves looks very much like the other, but the main difference lies in what type of voltage is being plotted against which type of current. In the common base configuration, the voltage across the collector-base junction is plotted against the current through the collector, or V_{BB} versus I_C. In the common emitter configuration the voltage across the emitter-collector circuit, V_{CE}, is plotted against I_C and in the common collector configuration the voltage across the collector-

67

emitter circuit can be plotted against I_E. In practice however, when looking at a manufacturer's specifications for a particular transistor, the V-I curve of the common emitter configuration usually yields all of the necessary information needed on a transistor.

Common Emitter V-I Curves

The method in which the family of characteristic curves is determined is by using a common emitter configuration and measuring values of I_C for various values of I_B. At the same time the value of VCE is also recorded. This performance circuit is shown in Fig. 4-14.

Notice in this circuit diagram the use of the ground symbol. This symbol is used more often than simply hard wiring all of the negative sides of the power supplies together. In this circuit, R_{adj1} is adjusted to supply various base currents. Variable resistor R_{adj2} is adjusted to give several values of VCE. By varying these potentiometers, changes in I_C can be noted and plotted on a V-I graph for the common emitter transistor circuit. A set of V-I characteristic curves is shown in Fig. 4-15.

The I_B in this family of curves has been set from 0 to 200 microamperes. R_{adj1} is varied to obtain each of the base currents desired. VCE is then varied over a range of from 0 volts to 50 volts while the transistor collector current, I_C, is recorded at each value of VCE. Plotting the relative VCE and I_C for different values of I_B results in the family of V-I curves shown above. Therefore I_C can be plotted when VCE equals 10 volts and when I_B equals 20 microamperes. This is point A. I_C can then be plotted again when VCE equals 10 volts but this time when I_B has changed or

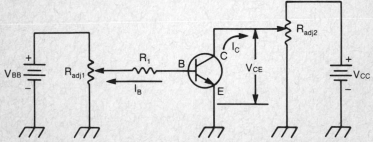

Fig. 4-14. A circuit used to determine the performance of a common emitter NPN amplifier circuit.

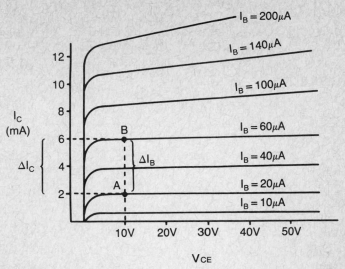

Fig. 4-15. The V-I curve for a common emitter circuit configuration showing the change in I_C for a change in I_B when V_{CE} is a constant 10 volts.

increased to 60 microamperes. This is point B. Looking at this more closely you can see that for a 40 microampere increase in I_B, I_C has increased by about 4 milliamperes. This increase is called current gain and is a measure of the level of how much the common emitter circuit will amplify current.

Common Emitter Current Gain

The current gain of a transistor in the common emitter configuration is given the Greek letter beta and is identified as β. Current gain can be found from the following equation:

$$\beta = \frac{\Delta I_C}{\Delta I_B}$$

β = current gain
I_C = change in collector current
I_B = change in base current

(while V_{CE} is held constant)

Using the previous example of changes in collector current with changes in base current, from Fig. 4-15:

69

$$\beta = \frac{\Delta I_C}{\Delta I_B} = \frac{6 \text{ mA} - 2 \text{ mA}}{60 \ \mu\text{A} - 20 \ \mu\text{A}}$$

$$\beta = \frac{4\text{mA}}{40 \ \mu\text{A}} = 100$$

With VCE at a constant voltage of 10 volts and a change in collector current of 4 milliamperes with a change in base current of 40 microamperes, this transistor has a gain of 100. This is a typical gain for low power bipolar common emitter transistors. There are, however, many low power transistors with much smaller gains and some low power transistors with gains considerably higher.

This amplification factor is also sometimes called the transistor's forward current transfer ratio, or h_{fe}, and is stated as the ability of the transistor to amplify a changing input current.

Common Base Current Gain

Although current gain can be found from the characteristic curve of a common emitter configuration, this is the ac current gain only. Current gain can also be found by using the common base configuration. In this case, the current gain is always near unity or slightly less than one because of the characteristics of this type of circuit configuration, where I_C is always less than I_E. Even so, manufacturers specify this current gain because it is the ability of the transistor to amplify a dc current rather than an ac current. This dc current gain is referred to as the dc beta and is identified by the symbol h_{FE}. It is the transistor's forward current transfer ratio and is given the Greek letter alpha, or α. Alpha is found with the equation:

$$\alpha = \frac{\Delta I_C}{\Delta I_E}$$

α = dc current gain
ΔI_C = change in collector current
ΔI_E = change in emitter current

There is a relationship between beta and alpha, stated as follows:

$$\beta = \frac{\alpha}{1-\alpha}$$

β = ac current gain of transistor
α = dc current gain of transistor

If the dc current gain of the transistor is known, the ac current gain of the transistor can be determined. As an example, if the alpha of a transistor is given as 0.97, the beta can be determined as follows:

$$\beta = \frac{\alpha}{1-\alpha} = \frac{0.97}{1-0.97} = 32.3$$

In a like manner, the equation can be transposed to find the transistor's alpha when the beta is known, as shown in the following equation:

$$\alpha = \frac{\beta}{\beta+1}$$

For example, given a beta of 100, a typical gain of a low power transistor, alpha can be found using the following method:

$$\alpha = \frac{\beta}{\beta+1} = \frac{100}{100+1} = 0.99$$

THE DC LOAD LINE AND Q POINT

One of the best ways to view the operation of a transistor under actual operating conditions is using a dc load line. With a specific load resistance, I_C can be viewed across varying collector-to-emitter volts, VCE, at different points of I_B, shown in Fig. 4-16. How a transistor works with a load is important in considering circuit design using a particular transistor.

Point A of the load line is usually VCC. Point B of the load line

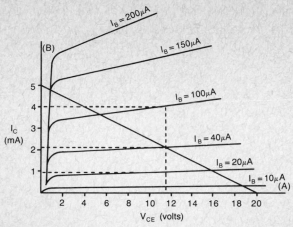

Fig. 4-16. The dc load line of a transistor shown here in general terms can be a very effective tool in determining a transistor's performance.

is usually IC_{max}. A line is drawn from point A to point B crossing the various values of I_B. In this case, point A is at a collector voltage of 20 volts, while point B is at a collector current of 5 milliamperes. With no signal input applied—that is, with the transistor at a resting condition—the base current is called the quiescent current and is found by plotting a point halfway up the dc load line. This is called the Q point. With the base current at this level, or 40 microamperes, I_C flows at a steady current of approximately 2 milliamperes. Also, about 12 volts dc exists across the collector and emitter of the transistor. The Q point in this example is set so that the amplifier operates with minimum distortion when used as an amplifier in the common emitter configuration. The dc load line serves to first establish the Q point of the transistor during the period when no signal is applied to the input. Secondly, the gain of the transistor can be found simply by finding the change in collector current for a change in base current. The dc load line is a more accurate way of measuring these differences than by viewing a transistor's V-I curve. As an example, look at Fig. 4-16. The collector current at a V_{CE} of 12 volts changes from 0.75 milliamperes to 4 milliamperes when the base current changes from 20 microamperes to 100 microamperes. The dc gain, or beta, of the transistor is therefore:

$$\beta = \frac{\Delta I_C}{\Delta I_B} = \frac{4 - 0.75}{.100 - .020} = \frac{3.25}{.08} = 40.6$$

MAXIMUM TRANSISTOR RATINGS

There are maximum ratings specified by the manufacturer for each type of transistor produced to protect the device from damage due to excessive voltages, currents, and/or temperatures. These specifications should be adhered to closely so that the transistor operates optimally. The following ratings are maximum ratings and are the safe allowable ratings that should be understood by anyone designing or replacing a transistor in an electronics circuit.

Breakdown Voltages

There are actually two breakdown voltages that you should be aware of when dealing with transistors: the collector-to-base reverse bias voltage, or collector breakdown voltage; and the emitter-to-base reverse bias voltage, commonly referred to as the emitter breakdown voltage. In the case of the collector breakdown voltage, too high a reverse bias voltage on the collector region of the transistor could cause the collector-base junction to break down, increasing the current flow to a very high level quite rapidly through the transistor and damaging it. To determine the maximum collector breakdown voltage, a reverse bias voltage is applied to the collector-base junction with the emitter lead disconnected, so I_E will be equal to zero. The collector breakdown voltage is therefore designated V_{CB0}. Also, a reverse leakage current is specified.

In determining the emitter breakdown voltage, the emitter-base junction is reverse biased with the collector lead open or disconnected, so I_C is equal to zero. But again, a reverse leakage current is specified. This breakdown voltage is termed V_{EB0}. Neither V_{CB0} nor V_{EB0} should be exceeded or the transistor will probably be damaged. When the manufacturer tests a transistor to determine these parameters the voltages are usually pulsed, rather than steady state dc voltages, so that the transistor is not damaged.

Another breakdown voltage specification listed by the manufacturer is the transistor's V_{CE0}. This is the transistor's collector-to-emitter reverse breakdown voltage and is measured with the base lead open.

In addition to these breakdown voltages the manufacturer also sometimes lists the maximum collector current, I_C, and emitter current, I_E, that the transistor can safely handle.

Collector Dissipation

Manufacturers also specify the maximum allowable power dissipation, or power rating, of the transistor. The power that a transistor dissipates is in the form of heat that comes from the reverse biased collector-base junction. It can be calculated by multiplying the collector current by the collector-to-emitter voltage. That is:

$$P = V_{CE} \times I_C$$

In practical terms, as the temperature of the transistor increases while it is conducting current, the power rating decreases. Usually, once the operating temperature of the transistor exceeds 25 degrees centigrade, a derating factor must be used to determine the true power rating of the transistor at various temperatures. The manufacturer usually supplies the information in the form of a graph. Where the transistor must operate in a high ambient temperature or when carrying large amounts of current, a heat sink is generally used to carry heat away from the transistor's junction. The heat sink, attached to the device, radiates the heat and helps protect the transistor from damage.

TESTING BIPOLAR TRANSISTORS

Testing a bipolar transistor is as easy as testing an ordinary PN junction diode. The difference is that there are now two junctions to test instead of just one. If you think of the transistor in the manner shown in Fig. 4-17, you should have no trouble in testing a bipolar transistor. In all cases, the forward resistance is measured using the R × 100 scale, while the reverse resistance is measured using the R × 10,000 scale.

NPN Transistor Testing

Testing an NPN transistor is simply a matter of using an ohmmeter and checking first the collector-base junction, then checking the emitter-base junction. Figure 4-18 shows how to properly set up the test procedure for an NPN transistor.

The ohmmeter is connected with the positive lead on the base and the negative lead on the collector. The resistance should be a low reading of approximately 500 ohms. Next, the negative lead of the ohmmeter is placed on the emitter lead of the transistor while

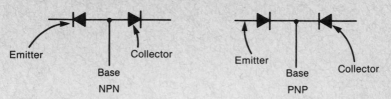

Fig. 4-17. For testing purposes, the NPN and PNP bipolar transistors can be viewed as having two junctions. Each junction can be tested just as with an ordinary PN junction of a diode.

keeping the positive lead on the base. In both cases this forward biases the two junctions of the transistor. Again, the ohmmeter should give a reading of approximately 500 ohms.

The next step is to reverse the ohmmeter leads and check the resistance of the two transistor junctions once again. This time the resistances should be very high, in the range of 700 k ohms or higher, because the junctions are now being reverse biased.

PNP Transistor Testing

This test setup is the same as that just described for testing NPN transistors, except that the ohmmeter is reversed in each of the steps. However, the readings should be the same.

Keep in mind that what is being tested here is not really a specific resistance measurement, but rather a forward to reverse resistance ratio which should be at least 10 to 1, just as with the PN junction diode.

In general, if both the forward and reverse resistance readings are very low, the junction is probably shorted. In contrast, if both the forward and reverse resistance readings of a junction are extremely high, then that junction is probably open.

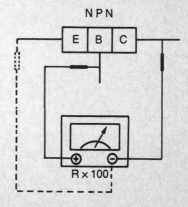

Fig. 4-18. This is how to properly test the two junctions of an NPN transistor in the forward biased mode with a reading of about 500 ohms being normal.

There are some transistor testers on the market that are used to measure a transistor either in a circuit or out of a circuit. Some are very good at measuring a transistor in a circuit, but it is usually recommended that a transistor be removed for the most accurate measurements. This is because the junctions could be across a very low resistance giving the indication that it may be shorted since it would show a low resistance reading on the ohmmeter in either direction. There are also transistor testers that measure the dc current gain of the transistor. Several companies manufacture ohmmeters that test transistors and display the current gain on a large LCD (liquid crystal display) screen. Not only can you determine if the transistor is either good or bad, but the gain is given as well.

Evaluating Unknown Devices

It is not uncommon to find a transistor that has no markings on it as to which lead is the emitter, base, or collector; or one that cannot be found in a suitable cross-reference guide to determine if it is a PNP or an NPN transistor. There is a method for making this determination using an ohmmeter. First, make sure that the positive voltage is on the red lead of the ohmmeter and that the black lead carries the negative voltage. This was discussed at length earlier when you were introduced to testing the PN junction diode in Chapter 2.

If possible, identify the base and emitter leads. Then connect either lead of the meter to the base lead of the transistor. The other lead goes to the emitter. Now check the resistance reading. If it is low, approximately 500 ohms, the junction is forward biased. A very high reading of nearly one megohm tells you that the junction is reverse biased. By observing the polarity of the emitter lead, and using Table 4-1, the transistor can be identified as either PNP or NPN.

Table 4-1.

Ohmmeter reading	Negative lead to emitter	Positive lead to emitter
high	PNP	NPN
low	NPN	PNP

Chapter 5

Field Effect Transistors

The field effect transistor, or FET as it is sometimes called, does not function like the bipolar transistor, although it is a three-terminal device like the bipolar transistor and can provide amplification as well. In many applications it is more suitable than the bipolar transistor.

Two types of FETs are discussed in this chapter. The first is the junction field effect transistor (JFET) and the second is the insulated gate field effect transistor (IGFET), also referred to as the metal-oxide semiconductor field effect transistor or MOSFET. Don't let these seemingly complicated terms fool you. Each of these field effect transistors is quite easy to understand.

CONSTRUCTION

It is necessary first to understand the construction of the junction field effect transistor in order to understand the theory of operation behind the device. As shown in Fig. 5-1, the JFET is constructed by lightly doping a semiconductor material called a substrate. This can be either an N-type semiconductor material or P-type. Next, an oppositely-doped region is formed within the substrate using a combination of epitaxial growth and diffusion. This process forms a channel within the substrate. If the substrate is N-type material, it is an N channel JFET. A P-type substrate is called a P channel JFET.

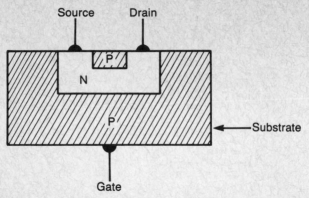

Fig. 5-1. A simplified illustration of the physical construction of a JFET.

Figure 5-1 shows that the oppositely doped material in the substrate has formed a channel so that it is flush with the upper surface of the substrate at two points. These points are called the source and the drain. Leads are connected to these two points. A third lead is connected to the substrate and is called the gate lead. Since the channel is symmetrically formed, the source and drain leads can be reversed without affecting the actual operation of the device. (This is not true for all JFETs.)

THEORY OF OPERATION

The diffusion of one type of semiconductor material into another causes a junction to be formed where the two semiconductor materials meet. This is true for JFETs as well as for PN junction diodes and bipolar transistors. And since current flows only when the junction is forward biased, the JFET needs biasing voltages for current to flow through it, as shown in Fig. 5-2.

In this case, a bias voltage is connected between the source and the drain of the JFET so that current flows through the channel of the device. This voltage is designated as V_{DS}. Here, an N channel JFET is being used as an illustration of JFET operation. The voltage connected between the gate and the source is used to control current flowing through the channel. It is called gate voltage, and is identified by the symbol V_{GS}.

Notice that V_{DS} is connected so that the source is negative with respect to the drain. Current flows through the channel due to the presence of majority carriers there. This current is referred to as drain current and is designated as I_D.

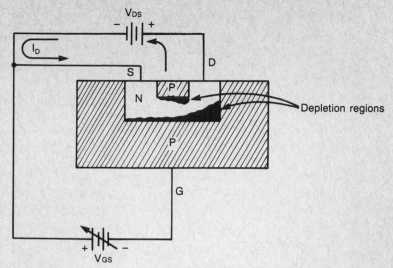

Fig. 5-2. A properly biased N channel JFET.

The method in which VGS is connected causes the gate to be negative with respect to the source, but this is a P-type gate and an N-type source. In effect, VGS reverse biases the junction formed by the P-type gate and the N-type channel. As with the ordinary PN junction diode, a depletion region forms around the PN junction of the gate and source. As VGS increases, so does the depletion region, effectively reducing the size of the N channel and thereby controlling current flowing through the channel. As VGS increases, I_D decreases, and as VGS decreases, I_D increases. Like the bipolar transistor, there is still a small leakage current flowing from gate to source. Also, the depletion region within the channel is somewhat larger near the drain end of the channel because VDS adds to VGS at the drain. It can then be stated that VGS actually controls the resistance of the N channel.

In contrast to a bipolar transistor, it is an input voltage that controls an output current rather than an input current controlling an output current. This can easily be seen by connecting an input signal between the gate and source and an output load between drain and source through which I_D can flow.

Also, the bipolar transistor, with its forward biased emitter-base junction, has a very low input resistance. The JFET has an extremely high input resistance due to the reverse biased gate-to-source junction.

If the gate-to-source voltage increases to a point where I_D is reduced to zero (or nearly so) this voltage is referred to as the gate-to-source cutoff voltage or $V_{GS_{(off)}}$. This is a specification usually listed on the manufacturer's data sheet.

Another factor to consider is the action that takes place within the channel near the drain as V_{DS} is increased in value. Up to a certain point, I_D increases also. But because V_{DS} effectively adds to V_{GS} near the drain end of the channel, increasing V_{DS} only increases I_D to a certain point. The depletion region begins to get larger near the drain end of the channel causing I_D to increase at a slower rate as V_{DS} is increased. Eventually, a point is reached where I_D is pinched off by the increasing depletion region. This specification is also provided by the manufacturer and is referenced with V_{GS} at zero volts. This pinch-off voltage is designated V_P, and I_D is designated as I_{DSS}, when V_{GS} is equal to zero volts and actually represents the JFET's maximum drain current.

CHARACTERISTIC CURVES

The V-I curve of a JFET is drawn like a bipolar transistor. Figure 5-3 shows the same circuit for an N channel JFET, this time using the standard symbol for this type of JFET. Notice the biasing on the gate, which is negative, and the biasing between drain and source.

Figure 5-4 shows the V-I curve for the circuit of Fig. 5-3. When V_{GS} is at zero volts dc, approximately 4.2 milliamperes of current flow through the N channel. As V_{GS} increases (becomes more negative), I_D decreases in value. For this particular JFET, 4.2 milliamperes of current is the maximum current that flows when V_{GS} equals zero volts and is therefore considered I_{DSS} in this case. When this maximum value of I_D is obtained V_P can be determined

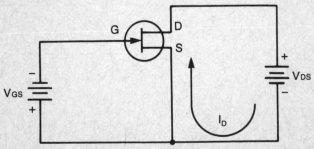

Fig. 5-3. The proper biasing scheme for an N channel JFET, shown using the symbol for this type device.

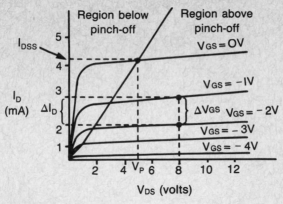

Fig. 5-4. The V-I curve for Fig. 5-3.

by dropping a vertical line down to VDS.

The operating region of the device below V_P is referred to as the ohmic region. Above V_P, the device is said to be operating in the pinch-off region. This is the region in which the JFET is normally operated. The JFET is said to be operating in the depletion mode since varying the depletion region controls the operation of the device.

The JFET achieves amplification by varying gate voltage above and below a set gate voltage in the pinch-off region. This set gate voltage is also called the JFET's Q point, just as in the case of the bipolar transistor. Thus the drain current, I_D, varies proportionately as an amplified representation of the input signal.

This gain in a JFET is called transconductance and is found with the following equation:

$$gm = \frac{\Delta I_D}{\Delta V_{GS}}$$

gm = transconductance in mhos
I_D = change in drain current
VGS = change in gate voltage

This equation holds true when VDS remains constant. As an example, if VGS changes from −1 volt to −2 volts, a corresponding change of I_D from 3 milliamperes to 2 milliamperes occurs. This is shown in Fig. 5-3. Use the equation for transconductance:

$$gm = \frac{\Delta I_D}{\Delta V_{GS}} = \frac{3\text{-}2}{3\text{-}1} = \frac{1 \text{ mA}}{1 \text{ volt}}$$

$$gm = \frac{.001}{1} = .001 \text{ mhos}$$

JFET Biasing

Just as it was appropriate to bias NPN and PNP transistors for proper operation, it is also essential to properly bias N channel and P channel JFETs for desired operation. Thus far the N channel JFET has been discussed and its biasing scheme has been shown in Fig. 5-3. The P channel JFET is biased in a similar manner, shown in Fig. 5-5.

In this biasing arrangement, V_{GS} and V_{DS} have been reversed. Their polarities are opposite those for biasing the N channel JFET. The arrow on the gate lead points outward; the arrow on the gate lead of the N channel JFET pointed inward.

Remember that in dealing with N channel JFETs, the majority current carriers are electrons, while in the P channel JFET, the majority current carriers are holes. Therefore, in the P channel JFET, the majority current carriers move from the source to the drain. The electrons, the minority carriers in this case, move from drain to source as shown in Fig. 5-5.

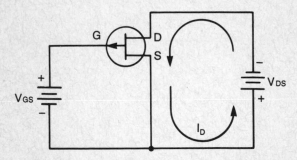

Fig. 5-5. Biasing for the P channel JFET is shown here with the arrow inside the symbol pointing outward.

MOSFETS

Another type of FET has similar characteristics to that of the JFET, but has no gate channel PN junction that must be reverse biased for proper operation. This is the insulated gate FET, or IGFET. It is also referred to as a metal-oxide semiconductor FET, or MOSFET.

The MOSFET can also be operated in two distinct modes. These are the depletion mode like the JFET, and the enhancement mode, which actually widens the channel rather than making it more narrow.

CONSTRUCTION

Construction of a MOSFET depends upon whether an enhancement mode or a depletion mode device is desired. Also, is the device to be an N channel type or a P channel type? There are N channel depletion mode MOSFETs, N channel enhancement mode MOSFETs, P channel depletion mode MOSFETs, and P channel enhancement mode MOSFETs. Each serves a particular function and all are constructed using similar methods.

The discussion of the construction of the N channel depletion mode MOSFET is given first. The P channel depletion mode MOSFET is similar; its discussion is limited to its biasing scheme. Next, a discussion on the N channel enhancement mode device is presented. Again, a discussion of the P channel enhancement mode is limited to a presentation of its biasing arrangement, since construction is similar to its N channel counterpart.

N Channel Depletion Mode MOSFET

The insulated gate FET is called that because the gate is electrically insulated from the semiconductor channel by using a thin silicon dioxide insulating layer, shown in Fig. 5-6.

This simplified drawing shows an N-type channel diffused into a P-type substrate. A thin layer of silicon dioxide is then placed above the substrate and consequently, above the N channel. Both ends of the thin layer are left open above the N channel material so that a source lead and a drain lead can be connected to the device. Another thin layer of metal is placed on top of the silicon dioxide causing the silicon dioxide to act as an insulator between the metal gate and the P-type substrate. These three layers of metal, oxide, and semiconductor give this device the name metal oxide semiconductor, or MOSFET.

This device is biased the same way an N channel JFET is biased, with the drain made positive with respect to the source and a controlling negative voltage on the gate. In the same way, an

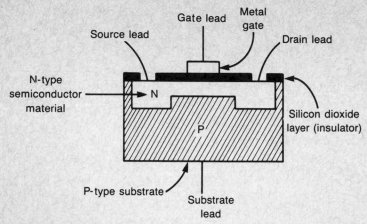

Fig. 5-6. Simplified construction of an N channel depletion mode IGFET also called a MOSFET.

increasing negative gate voltage produces a field effect that enlarges the depletion zone in the N channel. This decreases the flow of majority carriers (electrons), thus reducing I_D. The difference between the JFET and the MOSFET is that VGS can not only be increased to cause I_D to go to zero, but VGS can also be made positive to enhance the flow of electrons rather than depleting the N channel of these electrons. A positive VGS in an ordinary JFET cannot be allowed, but in the MOSFET the insulating layer prevents current from flowing into the gate. In effect, a negative VGS decreases I_D while a positive VGS increases I_D.

Since this device depletes the N channel of majority carriers when operated with a negative gate voltage, it is said to be operating in its depletion mode, or to be an N channel depletion mode device. When VGS equals zero, I_D is sufficient to cause this device to be normally on. This is true for all depletion mode devices, whether JFET or MOSFET.

Figure 5-7 is an illustration of an N channel depletion mode MOSFET that shows the proper symbol and biasing for this type of device. Notice the gate, drain, and source leads. There is another lead designated as SS or B. This is the substrate or base lead. This lead points inward for N channel devices as it did for the N channel JFET. The SS connection is usually an internal connection and is sometimes not shown. To understand the relationships between gate, drain, and source, you can view the MOSFET's drain characteristic curve, shown in Fig. 5-8.

84

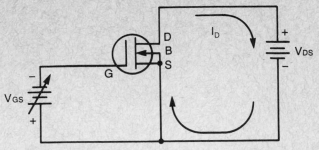

Fig. 5-7. The schematic symbol is shown here for this biasing scheme of an N channel depletion mode MOSFET.

These curves are similar to those of Fig. 5-4 for the JFET. The difference here is that positive as well as negative values for VGS are plotted on this graph, because VGS can be made either positive or negative. With the device operated below V_P, or pinch-off, I_D varies at a rather linear rate and over a large range as VDS varies, making this device useful as a voltage controlled resistor.

P Channel Depletion Mode MOSFET

The biggest difference between an N channel and a P channel depletion mode MOSFET is that in the latter type of device the majority carriers are holes instead of electrons, and these flow through a P channel instead of an N channel. Therefore, the

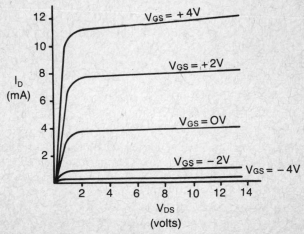

Fig. 5-8. The drain characteristic curve for the MOSFET of Fig. 5-7.

substrate in a P channel depletion mode device is a N-type semiconductor material while the drain and source are P-type semiconductor material. Although the drain to source voltage is opposite to that in the N channel device, the gate can still be either positive or negative.

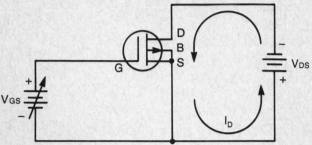

Fig. 5-9. Proper biasing and schematic symbol for the P channel depletion mode MOSFET.

The P channel depletion mode MOSFET biasing scheme and symbol are shown in Fig. 5-9. The arrow is pointing outward instead of inward and that V_DS has been reversed.

There is also a dual gate MOSFET. This symbol is shown in Fig. 5-10. This device is widely used as a gain controlled amplifier in broadcasting receivers. Notice the addition of zener diodes connected between gate and source leads, used to protect the device against static electricity that can break down the insulating layer at 35 to 50 volts. Static accumulates because gate resistance is extremely high, causing a capacitor effect between gate and substrate. This static can discharge and actually puncture the insulating layer.

N Channel Enhancement Mode MOSFET

There are many similarities between the N channel enhancement mode MOSFET and the N channel depletion mode MOSFET. However, the enhancement mode device does not have an N channel (or P channel) diffused into a substrate semiconductor material and is sometimes referred to as an induced-channel FET. This device is particularly suited for switching applications. Figure 5-11 shows the construction of such a device.

Since no conducting channel exists between drain and source, this device operates with no drain current when gate voltage equals zero. Usually, a zero gate voltage causes a minimum of majority carrier depletion within the channel and thus a substantial drain

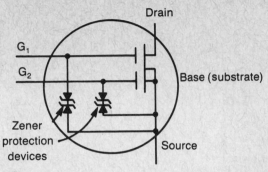

Fig. 5-10. Dual gate MOSFETs—not uncommon—represented as shown here.

current. In the N channel enhancement mode device, if the gate is made positive, it produces an electric field that attracts minority electrons in the P-type substrate. These electrons act as carriers and actually induce a channel from source to drain in the P-type substrate just under the gate. If the gate voltage increases to an even more positive value, I_D also increases, or is enhanced further.

Making the gate voltage negative does not affect the operation of the MOSFET since the drain current is zero anyway with no gate voltage applied. This means that the induced-channel MOSFET can only operate in the enhancement mode.

The difference in the depletion mode and enhancement mode devices can also be seen in their symbols. Figure 5-12 shows the symbol for the N channel enhancement mode MOSFET. In this

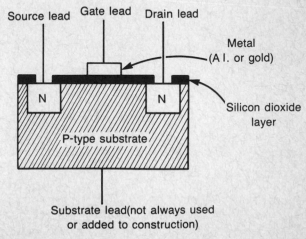

Fig. 5-11. Simplified construction of an N channel enhancement mode MOSFET.

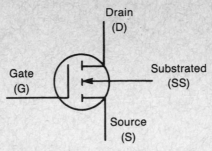

Fig. 5-12. The schematic symbol used to represent an N channel enhancement mode MOSFET (IGFET).

symbol, an interrupted line connects source, gate, and drain. This identifies the device as a normally-off MOSFET which is the condition of the device with no gate voltage applied.

Figure 5-13 shows a properly biased N channel enhancement mode MOSFET. Here, the drain is made positive with respect to the source while the gate is a variable positive voltage. Drain current flows only when V_{GS} is made positive. Again, drain characteristic curves show the relationship among source, gate, and drain in the N channel enhancement mode MOSFET, shown in Fig. 5-14.

With no positive gate voltage applied, no channel exists between drain and source. Once gate voltage reaches a threshold of about plus one volt, drain current begins to flow. Prior to this, the resistance from source to drain may be as high as 500 k ohms. Once V_{GS} reaches about 5 volts, this resistance decreases to about 3 k ohms.

Since a threshold is involved in turning on this device, this makes it suitable as a switch in digital applications. The area below this threshold level provides a region of noise immunity which

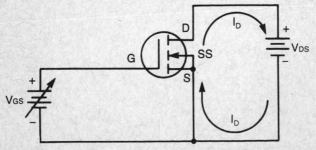

Fig. 5-13. An N channel enhancement mode MOSFET in its proper biasing configuration.

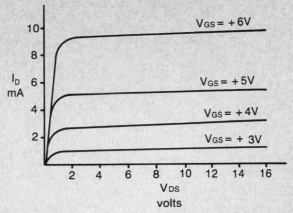

Fig. 5-14. A typical drain V-I curve for an N channel enhancement mode MOSFET.

prevents false triggering of the device to the on state. Therefore, it is normally either on or off, very much like a switch.

P Channel Enhancement Mode MOSFET

In the P channel device, the gate voltage must be made negative instead of positive because the minority carriers in the N substrate that must be attracted to the gate to form a channel are holes instead of electrons. The biasing scheme of the P channel device requires that VGS and VDS be just the opposite from those of the N channel device. However, the device is operated in the same manner as the N channel enhancement mode MOSFET, shown in Fig. 5-15, which includes both the proper biasing and schematic symbol for the P channel enhancement mode MOSFET.

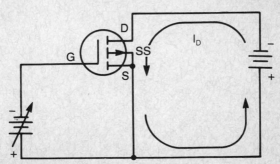

Fig. 5-15. The proper biasing and schematic symbol for a P channel enhancement mode MOSFET.

CMOS

If both N channel and P channel MOSFETs are formed on one chip, the combination of these two devices is called complementary MOS, or CMOS for short. Figure 5-16 is an example of a CMOS circuit that is used as a switch called an inverter because a high level input voltage is inverted into a low level output voltage, and a low level input voltage is inverted into a high level output voltage.

This is possible because in an N channel enhancement mode MOSFET (as used here) does not turn on unless its gate is a few volts more positive than its source. And a P channel enhancement mode MOSFET (again used here) does not turn on unless its gate is a few volts more negative than its source. The N channel MOSFET turns on and the P channel turns off when the input is near V_{DD}, connecting the output to ground. Therefore a high level input voltage causes a low level output voltage. When the input is near zero volts, the P channel MOSFET is turned on and the N channel MOSFET turns off, connecting the output to V_{DD}. A low level input voltage causes a high level output voltage. In each case the output is an inverted state of the input; thus, this circuit is called an inverter.

HANDLING FETs

Since the insulating layer within a MOSFET is susceptible to static damage, it is especially important to heed the manufacturer's

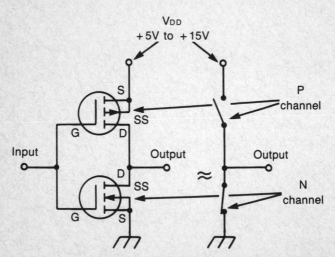

Fig. 5-16. The correlation between MOSFETs (used in a CMOS configuration) and switches.

specifications concerning maximum gate-to-source voltage, VGS. Even carrying a MOSFET without proper precautions can damage the device; for example, it can be damaged while you hold the device before placing it in a circuit. For this reason, manufacturers of printed circuit boards containing MOSFETs only allow those persons suitably grounded to handle these devices. This usually means that person wears a wrist strap attached to a wire that is connected to ground through a one megohm resistor.

Also, MOSFETs are normally stored in static foam that prevents static buildup between the leads. In addition, soldering MOSFETs into a circuit requires the use of a soldering iron that is properly grounded. Finally, MOSFETs are never inserted or removed from a circuit while the power is on. One final note: MOSFETs can also be damaged if handled by persons wearing clothing that builds up a large static charge on a dry day. This includes clothing made from synthetic fiber such as nylon.

TESTING FETs

The MOSFET has been described as a device that can be damaged merely by handling it without following the proper grounding techniques. The best way to test an FET, whether JFET or MOSFET, is by testing it while it is operating in a circuit. This requires knowing what type of FET you will be testing so that the correct voltages, positive and negative, can be established prior to making measurements.

The N channel JFET or N channel depletion mode MOSFET are biased in the same manner; that is, the drain is made more positive with respect to the source while a controlling negative voltage is applied to the gate. In the N channel enhancement mode MOSFET, you would not expect to find a negative voltage on the gate lead since it is a positive gate voltage, on this particular device, that causes drain current to flow. In essence then, testing any FET device requires first knowing the type of device that is to be tested and secondly, understanding what correct voltages are necessary for a properly operating device.

You may want to test devices on the work bench either because the device is not or will not soon be placed into an active circuit or because you have several different types of solid-state devices and would like to compare the testing techniques that are performed on each. In that case, you can record readings on an ohmmeter when testing JFETs to determine the value of the device. MOSFETs are better tested in-circuit. Figure 5-17 shows an N channel JFET and

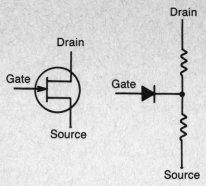

Fig. 5-17. An N channel JFET and its equivalent circuit.

its equivalent internal construction in terms of resistance. Remember, this is an equivalent circuit only, meaning that the device acts in an equivalent manner when current is passed through it from the ohmmeter.

Whether measuring from drain to source or source to drain, the resistances are the same and do not depend on the polarity of the ohmmeter leads when performing this test. This is a fixed resistance of from 100 ohms to 10 k ohms. Also the gate to source and gate to drain resistances are identical. When forward biasing the source or drain junctions, the ohmmeter should read approximately 1 k ohms. And in reverse biasing these junctions, they should appear to be open because of the high impedance characteristics of the gate.

Figure 5-18 is a P channel JFET and its equivalent circuit. Again, the drain to source and source to drain resistance readings are the same. And, as with the N channel JFET, forward biasing

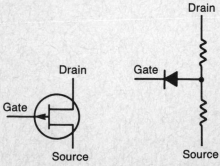

Fig. 5-18. A P channel JFET and its equivalent circuit.

the gate to drain or gate to source junctions causes a relatively low resistance reading of 1 k ohm, while reverse biasing these junctions causes the junctions to appear to be open.

BASIC FET CIRCUITS

There are a number of ways in which an FET can be connected or configured in an electronic circuit. Just as the NPN and PNP bipolar transistors were configured differently, depending on their use, FETs can also be configured differently depending on their use. The three circuit arrangements discussed in this section are the common source, common gate, and common drain. These circuit arrangements are valid for both JFETs and MOSFETs and one lead is common in each of these circuit configurations.

Common Source (Inverting Amplifier)

The common source circuit is probably the most popular of the three circuit arrangements, and is shown in Fig. 5-19. In this circuit, the source lead is common between input and output signal paths. The input is applied between source and gate while the output is taken across the source and drain.

In this circuit, R_G is effectively in parallel with the gate-to-source resistance when an input signal is applied and must be a high value to prevent lowering the input resistance of the JFET. The input signal always sees a high input resistance. The device is operated within its pinch-off region by adjusting V_{GG} for a set value of drain current, I_D. R_L develops the output voltage signal since I_D flows through it causing a voltage drop. V_{DD} drops some

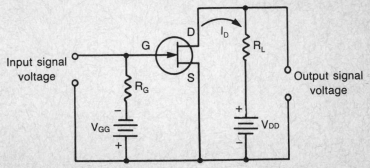

Fig. 5-19. A common source (inverting amplifier) circuit configuration using an N channel JFET.

93

of its voltage across R_L while the rest of it is dropped across the FET.

In the bipolar transistor, the input current controlled the output of the device. In this circuit it is the input voltage that controls the output, so voltage amplification takes place since a small change in input voltage causes a large change in output voltage. However, the output voltage signal is a reversed amplified version of the input signal, because as the negative gate voltage increases with the addition of the negative half cycle of the input signal, drain current increases. More voltage is now developed across R_L with less voltage developed across the drain to source of the JFET. The opposite occurs when V_{GS} goes less negative or to a more positive value. Thus, the output voltage signal is an inverted amplified version of the input signal voltage. For this reason the common source circuit can also be called an inverting amplifier.

Another feature of this amplifier is its relatively high input resistance, making it suitable in digital applications where input loading must be considered when connecting a single circuit output to several circuit inputs.

Common Drain (Non-inverting Amplifier)

The circuit configuration for the common drain circuit is shown in Fig. 5-20. Here, the input signal is applied between the gate and drain while the output signal is developed across the source and drain with the drain being common to both input and output signal paths. It may not appear at first glance that the drain lead is the

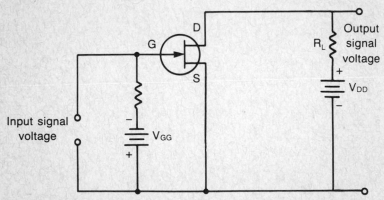

Fig. 5-20. The schematic for a common drain circuit configuration using an N channel JFET.

common lead. However, the dc voltage property of VDD is effectively a short as far as an ac input or output signal is concerned. In effect then, the drain lead is common to both input and output since VDD acts as a short to ground for the drain lead of the FET.

This circuit is also referred to as a source follower (like the emitter follower of the bipolar transistor) since the output voltage at the source follows or tracks the input voltage. And, as in the emitter follower, the voltage gain is less than unity or one.

Common Gate

The common gate circuit is another popular circuit configuration for the FET or MOSFET. An N channel JFET common gate circuit is shown in Fig. 5-21. In this circuit arrangement, the input appears between the gate and source while the output is developed across the gate and drain with the gate being common to both the input and output signals.

Instead of VGS controlling drain current, VSS, or source voltage, is used to vary drain current. Any variations in input voltage vary drain current and the voltage dropped or developed across R_L. These variations in output voltage follow those of the input signal voltage, so there is voltage amplification without signal inversion.

In this circuit arrangement, the common gate presents a relatively low input resistance and high output resistance making it suitable in applications where a high resistance load needs power supplied to it from a low resistance source. This circuit is also very useful in high frequency amplification, because of its stability due to lack of oscillations at high frequencies. This ability to be free of oscillations during amplification is apparent from the fact that

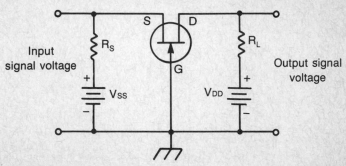

Fig. 5-21. A common gate circuit configuration using an N channel JFET.

the output is effectively isolated from its input, preventing feedback from output to input causing unwanted oscillations.

Any of the three above circuit configurations can be constructed using N or P channel enhancement mode or depletion mode MOSFETs or P channel JFETs. All that is needed is that the polarities of the bias voltages be properly connected for correct operation. And, when using MOSFETs, the lead marked as the substrate is usually connected to the circuit ground.

GaAsFETs

There are now competitive FET structures that replace standard MOS devices in the electronics industry. Some of these are the Metal-Nitride-Oxide-silicon (MNOS) device, the Resonant-Gate Transistor, and the Metal-Nitride-Gallium Arsenide FET, better known as the GaAsFET. To understand the operation of the GaAsFET requires understanding some of the properties of silicon dioxide and gallium arsenide that distinguish these two semiconductor materials from each other. The theory behind the operation of GaAsFETs does not depend on learning about an entirely new device, but rather on learning how GaAs affects the operation of devices already studied. As you will see, it has become the preferred substrate for both discrete devices and integrated circuits for a number of good reasons. But generally, it is the increased mobility of electron transfer through an FET constructed using GaAs that allows it to be far superior in operation where high speed switching is needed. This makes the GaAsFET desirable for high frequency applications and it is in microwave amplifiers that you may first encounter a GaAsFET.

Up to this point the term transconductance has not been used, but it is mentioned here so that you may better understand the performance of a GaAsFET. The symbol for transconductance is g_m and its definition states that for a constant value of drain-to-source voltage, V_{DS}, g_m is directly proportional to a change in drain current and inversely proportional to a change in gate-to-source voltage. The equation looks like this:

$$g_m = \frac{\Delta I_D}{\Delta V_{GS}}$$

with V_{DS} held constant.

Originally transconductance was measured in units called mhos, but that has been changed to the siemen or siemens. Notice that the term mho is the opposite spelling of the term ohm. Therefore since ohm is a unit of measure of the resistance to current flow, mho, now siemen, is the measure of the admittance of current flow through a semiconductor material. An increase in g_m then means that electrons and holes travel faster through devices with a high g_m.

In MOSFETs, N channel devices are preferred over P channel devices because electron mobility is about twice that of hole mobility. One reason GaAs is used is because it has an electron mobility that is about 10 times that of silicon. In effect then, GaAsFETs have a much higher g_m than the typical MOSFET constructed of silicon.

Initially, silicon dioxide was used as the gate dielectric in the first GaAsFETs, but this material did not work well with GaAs as the substrate. Performance was limited in many respects. Eventually silicon nitride replaced silicon dioxide as the dielectric and a high resistivity material was used as an epitaxial layer between the substrate and the dielectric, as shown in Fig. 5-22. As you can see, this structure is similar to that of a MOSFET device, but with a performance that has been much improved. This can be seen in Fig. 5-23. Power gain remains higher at higher frequencies for the GaAsFET than for the silicon device. Presently, some devices can be operated at frequencies approaching 20 GHz.

A typical GaAsFET characteristic curve is shown in Fig. 5-24. It is similar in content to any FET device with V_{DS} plotted against I_{DS}. If a constant V_{DS} is selected, g_m can be found by finding a

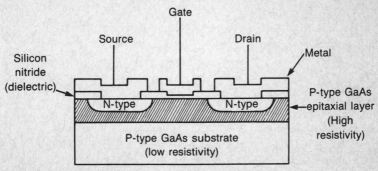

Fig. 5-22. Construction of a typical GaAsFET.

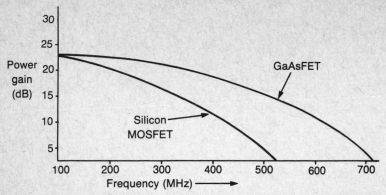

Fig. 5-23. Comparison between a silicon MOSFET and a gallium arsenide FET.
Notice the better power gain over a wider range of frequencies for the GaAsFET.

change in drain current and dividing that result by a change in gate-to-source voltage. In Fig. 5-24 if a VDS of 4 volts is chosen, g_m can be found by finding, as an example, a change in VGS of from -2.0 volts to -1.0 volts. This would cause a change in I_{DS} from about 0.6 amps to 1.2 amps. Therefore g_m could be calculated as follows:

$$g_m = \frac{\Delta I_{DS}}{\Delta VGS}$$

$$g_m = \frac{1.2 - 0.6}{-2 - (-1)}$$

$$g_m = \frac{0.6}{-1} = 600 \text{ mS}$$

This is a typical level of transconductance for a device used in high power (20 watts), high frequency applications. Incidentally, even though 0.6 amps is divided by a minus 1, the result of the equation is a positive number since moving from -2 to -1 results in moving from a negative number to a less negative number, or towards a more positive number.

In addition to a characteristic curve, many manufacturers have a curve indicating the relative power output of a device in comparison to the input power, because many applications require cascading devices, from low to medium to high power. It is essential that you understand the input and output limits of each device in the chain to prevent overdriving any or all stages. Overdriving

98

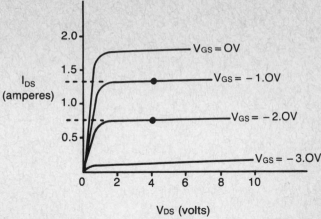

Fig. 5-24. A typical GaAsFET characteristic curve, I_{DS} versus V_{DS}.

causes a phenomenon which is called the gain compression point (G.C.P.) of the device. Effectively, when output power will increase no more than 1 dB despite a greater than 1 dB increase in input power, the device is said to have reached its maximum output power at 1 dB G.C.P. Manufacturers also specify this value using a graph, usually listing the G.C.P.s for several of their products.

Finally, transistor outlines for several GaAsFETs are shown in Fig. 5-25.

THYRISTORS: SPECIAL PURPOSE DEVICES

There are a number of other devices besides FETs that can be used to perform a dedicated function in an electronics circuit. Where the bipolar transistor or JFET may be limited in its abilities to perform certain functions, other devices, referred to as thyristors, can meet these needs. Thyristors are designed specifically for switching on and off, to either block current or to allow it to flow. By doing this, they can also regulate the amount of power within an electronic circuit. While it is true that bipolar transistors and FETs can be configured to also do these functions, thyristors are designed to handle considerably heavier current demands. In effect, thyristors are used where large power handling capabilities are required whereas transistors and JFETs are used primarily as amplifying devices.

The thyristors presented here are the silicon controlled rectifier (SCR), the triac, and the diac. Other thyristors are the unijunction transistor, or UJT, and the programmable UJT, or PUT. These

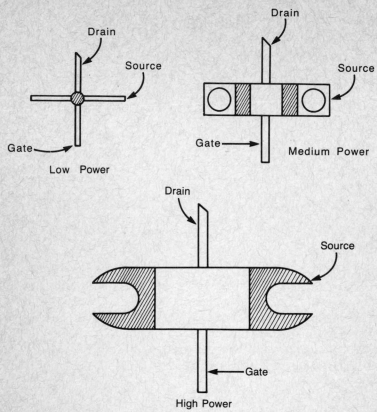

Fig. 5-25. Outlines for low, medium, and high power GaAsFETs. Mounting tabs for screws are used on the medium and high power devices.

last two devices are discussed further in Chapter 10, Oscillators, since they are more often suitable in that type of electronic application.

Silicon Controlled Rectifiers (SCRs)

The silicon controlled rectifier (SCR) is similar in its construction to the bipolar transistor but contains three PN junctions instead of two and is used almost exclusively as a switching device. Since it is by design a rectifier or diode, it conducts current in one direction only. The basic construction of an SCR is shown in Fig. 5-26. It is basically a PNPN semiconductor device with three leads labeled anode, cathode, and gate. These terms are actually holdovers from the vacuum thyratron tube which uses those terms

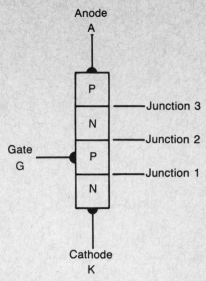

Fig. 5-26. Simplified construction of an SCR.

to identify its separate parts. In effect, the cathode is the emitter, the gate is the base, and the anode is the collector region of the SCR.

An equivalent transistor circuit for the SCR is shown in Fig. 5-27. Here, Q_1 is a PNP transistor while Q_2 is an NPN transistor.

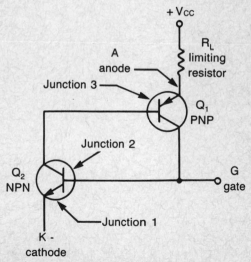

Fig. 5-27. Equivalent bipolar transistor circuit of an SCR.

The emitter of Q_1 is connected to VCC through a limiting resistor since once turned on, like a diode, it allows a short circuit path for current to flow. If the gate of the SCR is made more positive than the cathode, K, the emitter-base junction of Q_2 is forward biased allowing a negative potential on the base of Q_1. With the gate connected to the collector of Q_1, the collector-base junction of transistor Q_1 is now forward biased, turning it on also. A positive voltage on the gate lead of the SCR turns the SCR on (allows current to flow) as long as the cathode is negative with respect to the gate and anode, and the anode is positive with respect to the gate and cathode. Once the positive potential on the gate is removed, the SCR continues to conduct current like a diode or rectifier. Only by reducing the anode voltage to near zero or removing it completely will the device cease to allow current to flow. A single positive pulse on the gate of the SCR is all that is needed to turn this device on and keep it on. And as with any diode, reverse biasing causes a very small leakage current to flow.

The schematic symbol for a properly biased SCR is shown in Fig. 5-28. Here, V_G is made positive to turn the SCR on allowing forward current, I_F, or anode current, I_A, to flow. A more detailed analysis of the SCR can be shown in Fig. 5-29. Here, the current versus anode voltage characteristic curve is displayed. This curve is called the SCR's V-I curve. In this curve, the upper right quadrant is an indication of the effects of forward biasing the SCR without the use of a positive gate voltage. A steady increase in negative voltage is applied between cathode and anode until a point in forward is reached where the SCR breaks down and conducts current. The voltage at which this takes place is called the breakover voltage, VBR. SCRs usually are not operated at voltages

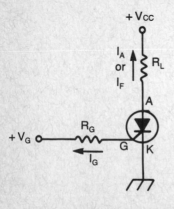

Fig. 5-28. Properly biased SCR using the SCR symbol.

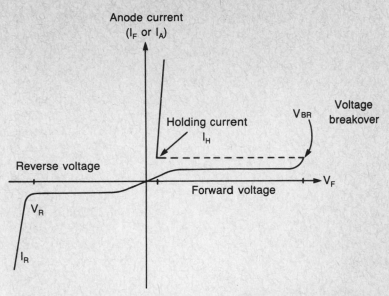

Fig. 5-29. V-I characteristic curve for a typical SCR with gate voltage left open.

near this value. However, once V_{BR} is reached, the only limiting factor for current flow is a limiting resistor in series with the device.

The SCR is now on and the forward voltage drop, V_F, immediately drops to a much lower value due to the reduction in resistance of the device once breakover occurs and current flows. Now only a small V_F is needed to produce a large forward current, and this large forward current continues to flow as long as I_F does not drop below a minimum current known as the holding current, I_H. This curve also shows that the SCR functions very much the same way that an ordinary PN junction diode functions when it is reverse biased, with no gate voltage or with the gate lead left open. There is also a relationship that exists between forward voltage, V_F, and forward current, I_F, when different values of positive voltage are applied to the gate, shown in Fig. 5-30. Here, as you can see, different gate voltages produce different forward voltage breakover points in the curve. As gate current increases, the forward breakover voltage point occurs earlier along the forward voltage, V_F, line of the graph. A relationship therefore exists between gate current and forward voltage applied to the SCR. Simply stated, for any given value of gate current, a forward breakover voltage must be reached before the device will turn on. And for any given value of forward voltage, a gate current must also be

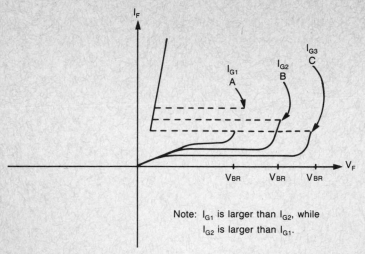

Note: I_{G1} is larger than I_{G2}, while
I_{G2} is larger than I_{G1}.

Fig. 5-30. Different values of gate current produce different values of breakover voltage.

reached before the device turns on. The operation of the SCR is a function of both gate current and forward voltage.

A typical application of an SCR is in the control of ac power to a load. This circuit is shown in Fig. 5-31, while its output waveforms are shown in Fig. 5-32. Remember, the SCR is basically a cousin to the diode and therefore conducts current in one direction only. Keeping this in mind, a 120 V 60 Hz signal is applied to the input of the circuit. As the negative alternation of the sine wave is applied, C_1 is charged through the forward biased diode

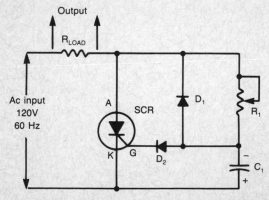

Fig. 5-31. A typical SCR circuit used as a half-wave phase controller with the amount of power to the load determined by the setting of R1.

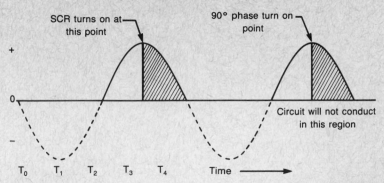

Fig. 5-32. The shaded area indicates the output portion of the input ac signal if R1 were increased to maximum value.

D_1. At this point the SCR does not conduct because the device is reverse biased on the negative alternation of the input ac signal. During the positive alternation of the input signal, the SCR is forward biased through R_L. Also, D_2 is now forward biased through R_L and R_1, applying a positive voltage to the gate of the SCR. This positive voltage is determined by R_1 and C_1. During this positive period, the capacitor discharges and then recharges again through R_1 instead of D_1, since D_1 is now reverse biased. The lower the value of R_1, the quicker C_1 charges to a positive value, and this positive voltage is applied to the gate of the SCR through D_2. Stated another way, the lower the value of R_1, the sooner the SCR turns on during the positive portion of the input ac signal. The SCR can be made to turn on during any portion of the positive alternation of the ac input signal by adjusting R_1. By controlling the phase of the ac voltage across a load, the ac power applied to the load is also controlled. This makes the SCR suitable to control the amount of power supplied to a load from zero to about 50 percent of the applied input ac power.

In effect then, when R_1 is equal to zero or at its minimum setting, the SCR in this circuit acts like an ordinary PN junction diode, conducting only during the positive portions of the ac input signal. Varying R_1 has the effect of causing the SCR to turn on during different times, varying the phase of the positive output signal. Increasing R_1 allows the SCR to turn on later during the positive phase of the input signal. The device can be turned on at the start of the positive input cycle, such as 30 into the start of the positive cycle, or all the way up to 90 into the positive input signal, controlling the power of the output signal. This is shown more graphically in Fig. 5-33.

105

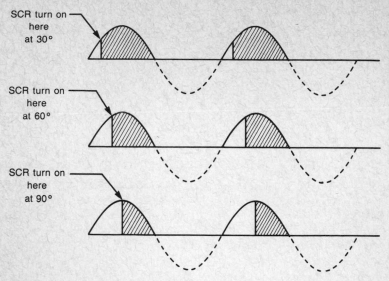

Fig. 5-33. The SCR can be turned on during almost any portion of the input ac waveform producing the output waveforms shown with the shaded areas indicating conduction.

Triacs

The SCR was useful for controlling power to a load. However, the output from an SCR circuit is a series of positive pulses, or certain portions of those pulses. An ideal circuit is one that uses both the positive and the negative cycles of the input ac signal. Since the SCR is a unidirectional device, meaning that current can flow through it in one direction only, a more useful device is one that is bidirectional, that allows current to flow through it in either direction, but again, in a controlled manner. This device is the triac, a bidirectional triode thyristor.

The triac basically has four major regions, NPNP, with two smaller regions located within the last P area. The construction of the triac is shown in Fig. 5-34. The triac has three connections designated as main terminal 1, main terminal 2, and gate. Sometimes the main terminals are just labeled as T_1 and T_2. Each of these leads, as can be seen, is connected to a PN junction. An equivalent SCR circuit is shown in Fig. 5-35. These two SCRs are not connected back-to-back, but rather in parallel, allowing current to flow in either direction. Keep in mind that this is an equivalent circuit only and not a true representation of the inside of a triac. The standard symbol used to identify the triac is shown in Fig. 5-36.

The V-I characteristic curve for a triac is shown in Fig. 5-37.

106

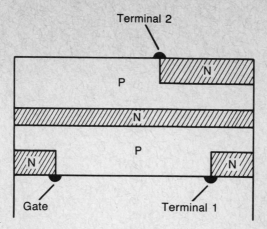

Fig. 5-34. Simplified construction of triac showing two main terminals and a gate.

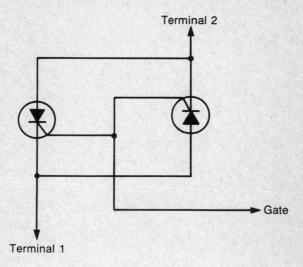

Fig. 5-35. Equivalent SCR circuit of a triac.

Fig. 5-36. Schematic symbol for the triac.

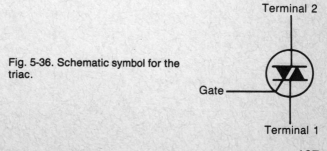

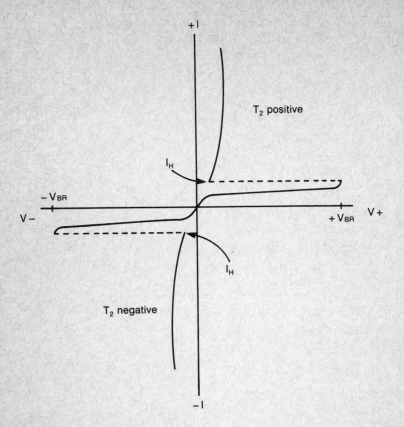

Fig. 5-37. Typical V-I curve for a triac.

This is a graph of anode current versus anode voltage, with no gate current flowing. In this case, T_1 or T_2 is the anode. When T_2 is made positive with respect to T_1, the curve is a duplicate of the V-I curve of a typical SCR. When T_2 is made negative with respect to T_1, the same curve is obtained but in the opposite direction. Since T_2 has been made negative, I_T in the reverse direction is shown as minus I_T, although negative current doesn't actually exist. The minus sign here merely indicates current flowing in the opposite direction.

As in the SCR, V_{BR} is really a function of gate current, I_G, and the triac turns on sooner as gate current is increased. However, the triac operates with either a positive or negative gate voltage. T_1 is used as the reference terminal and the triac's V_{BR}, either forward or reverse, can be reduced by making V_G either more positive or more negative with respect to T_1. The triac can be turned on by allowing a gate current to flow into the gate lead (a negative gate voltage) or by a positive gate voltage where gate current flows out of the gate lead. And, in turning off the triac, the operating current, $+I_T$ or $-I_T$, is simply reduced below the holding current, I_H.

Remember, in either case only a momentary gate voltage is required to turn the triac on. To determine what value the gate voltage should be, manufacturers of these devices specify typical and/or minimum values of gate current, both plus and minus, and positive and negative values of gate voltage required for proper operation of a particular device.

Because the triac is a bidirectional device, it is suitable for use as a full-wave phase control circuit device. Figure 5-38 is an illustration of how a triac might be used as a light dimmer. In this circuit, the input ac signal contains both positive and negative alternations. This causes C_1 to charge and discharge through R_1, the variable resistor. As the voltage across C_1 reaches the required level, the triac is turned on through a triggering device, a switching device that turns on when a certain voltage level is reached, and turns off when that same level is reduced. When the triggering device turns on, gate current flows momentarily and the capacitor discharges though the triac. Once the capacitor discharges enough, the triggering device turns off. The charging time of C_1, and consequently how quickly the triggering device turns on, is a function of R_1 since R_1 controls the charge time of the capacitor. With R_1 at maximum value, the capacitor takes longer to charge causing triggering to occur later in each ac alternation and thus the power applied to the lamps is reduced. Therefore R_1 controls

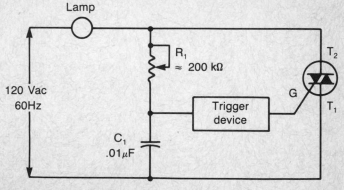

Fig. 5-38. A typical light dimming circuit using a triac.

the brightness of the lamp. Without drawing the output waveform across the lamp, you can merely picture the output waveform of the SCR and add the negative alternations on the bottom half of each cycle in the waveform. And, of course, the negative half cycles have their phase affected in the same way as the positive alternations.

One note about the triggering device: This is a diac, to be discussed next, which compensates for the nonlinearity of the triggering characteristics of the triac as opposed to the more stable triggering characteristics of the SCR. In essence, the triac must trigger on gate currents flowing in opposite directions while the SCR has only to contend with a unidirectional current.

Probably the biggest disadvantage of the triac is in its inability to handle very large amounts of current like its cousin, the SCR, can. An SCR can safely switch currents of up to 750 amperes or more, while triacs are usually limited to less than 50 amperes.

Diacs

The diac is constructed very much like the ordinary bipolar transistor and is very often used as a triggering device in triac circuits. Figure 5-39 shows the construction of a typical diac and its equivalent circuit. During the manufacturing process leads are attached only to the outer two semiconductor regions, while no lead is attached to the center region. Also, in the bipolar transistor, doping concentrations around the junctions are somewhat uneven. Here, in the diac, these concentrations are very nearly the same. Because of this type of doping, the diac exhibits the same effect on current in spite of the direction in which current may be flowing.

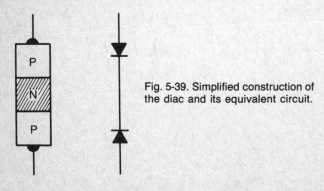

Fig. 5-39. Simplified construction of the diac and its equivalent circuit.

As you can see, the only current that this device allows to flow, while in its off state, is leakage current. In the equivalent circuit, no matter which diode junction is forward biased, the other is reverse biased. Obviously, current does not flow through the diac until one of the diode's junctions is broken down due to a high enough reverse bias voltage. Once this reverse bias voltage level is reached, current flows just as it would through a closed switch and is limited only by a resistance that is placed in series with the device. This is very similar to the triac where a certain value of breakover voltage, with no gate voltage applied, causes the device to conduct current. In fact, the V-I characteristic curves of Fig. 5-40 are very nearly the same as a typical triac.

This V-I curve shows that once a particular breakover voltage, V_{BO}, is reached in either direction, the diac begins to conduct current flow. Once current begins to flow, the internal resistance of the diac decreases causing a lower voltage drop across the device with a substantial increase in current flow. Most diacs break over at approximately 30 volts, although this value varies somewhat from one diac to another.

Earlier you saw that the triggering device used for turning on the triac was the diac now being discussed. That circuit is again represented in Fig. 5-41. In this illustration, the diac symbol is used as the triggering device. Again, as C_1 charges through R_1, the V_{BO} of the diac is reached, turning on the device and allowing current to flow through the triac's gate, turning that device on. C_1, however, discharges quickly through the diac, the gate of the triac, and

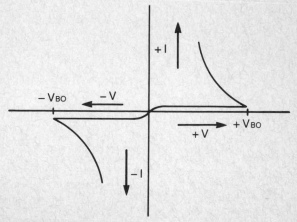

Fig. 5-40. The V-I curve of a typical diac.

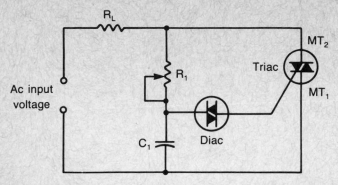

Fig. 5-41. A typical light (R_L) dimming circuit using a diac and a triac and can also be used for controlling the heating element of a soldering iron.

its terminal MT_1. The gate of the triac actually sees a pulse of current, not a steady state one. Since the internal resistance of the diac changes once VBO occurs, it is one of the factors that determines the level and duration of the current pulse. Remember too that the triac's gate to MT_1 resistance is also a part of this discharge circuit so it too, along with the size of C_1, determines the shape of that current pulse. Manufacturers give enough information in their specification sheets on diacs to give you the values needed to determine proper capacitance to use for a particular diac.

Chapter 6

Optoelectronic Devices

The solid-state devices discussed thus far amplify or switch current and voltages, and are constructed so that the control of these devices is accomplished, in most cases, directly. In other solid state devices, current flow is controlled indirectly. These devices either convert light energy into electrical energy or electrical energy into light energy and are called optoelectronic devices.

There are two classes or categories of optoelectronic devices: those that are light emitting and those that are light sensitive. There is also a device called an optical isolator that uses the characteristics of both types of devices. This chapter begins with an introduction to the principles of light including photons and the units in which light is measured.

BASIC PRINCIPLES OF LIGHT

The characteristics of light are very similar to those of other types of electromagnetic radiation; that is, light behaves in a manner that is consistent with such waves as radio frequency (rf) waves. However, unlike rf waves, electromagnetic radiation in the frequency spectrum between 300 gigahertz and 300 million gigahertz produces waves that are visible to the human eye. It is these waves that allow you to see the writing on this page, and the page itself.

Within the frequency spectrum of all light are an infinite number of other frequencies. These other frequencies of electromagnetic radiation give us all the colors we can see. Those colors range from red (the lowest visible frequency) to violet (the highest visible frequency). In fact, all of the colors of the rainbow that you may have seen during or after a rain shower are those colors contained within the frequency spectrum of light. Figure 6-1 is a simplified drawing of the frequency spectrum. The visible light spectrum takes up only a very small part of that range of electromagnetically radiated frequencies.

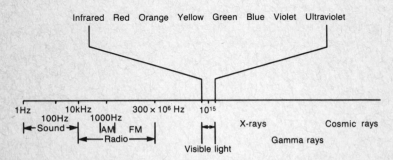

Fig. 6-1. A simplified representation of the electromagnetic spectrum.

At the lower end of the frequency range, visible light can be seen as the color red. Just below that frequency is the area of radiation known as infrared (below red). At the other end of the visible light range can be seen the color violet and just above this is the ultraviolet (above violet) frequency range of electromagnetic radiation. As you can see, light in the infrared range can be very useful because it still behaves as light, but is invisible to the human eye. This makes it suitable for such applications as burglar alarms and night security surveillance systems. (More about this when light emitters and light sensors are presented.)

The major difference between most other electromagnetic waves and light waves is that light waves appear to contain characteristics not found in the radiation in other parts of the frequency spectrum. Light definitely behaves very much like rf waves when passing through glass or water; that is, it is displaced and its velocity reduced. But other explanations must be given when light strikes a photosensitive device, such as certain types of semiconductor materials. These other characteristics of light can be explained by quantum theory. This idea or theory of light suggests that light not only has characteristics similar to those of other forms of electromagnetic radiation, but also behaves as though it is made up of very tiny invisible particles. Scientists call these invisible particles photons. Photons contain energy and are viewed as a quanta (packet) of energy.

Photons are actually rather interesting particles. They are charged neither negatively nor positively, but rather are neutral in charge. And, according to quantum theory, those photons at the higher end of the light spectrum contain more energy than those photons at the lower end of the spectrum. Violet light has more energy content than red light. In fact, all along the electromagnetic frequency spectrum those forms of electromagnetic radiation at the higher frequencies have more energy content than those forms of electromagnetic radiation at the lower end of the frequency spectrum. It is this quantum theory of light that is used to explain how semiconductors can act as light sensors and light emitters.

Since light waves are part of the frequency spectrum and are propagated (travel) through space like other electromagnetic waves, their wavelength is measurable just as is any other rf wave. And, of course, the same equation then is used in finding the wave length of light. Since the frequency of light spans a certain range, its wavelength varies also, from a minimum to a maximum. Finding the wavelength of light at any particular frequency then, requires the use of the following equation:

$$\lambda = \frac{V}{f}$$

λ = one complete wavelength of light (λ is the Greek letter lambda)
V = velocity of electromagnetic radiation, usually in miles per second or centimeters per second
f = frequency in hertz

The velocity of light in a vacuum is approximately 186,000 miles per second or 3×10 centimeters per second, for all practical purposes. This means that finding the wavelength of light at its highest frequency requires solving the following equation:

$$\lambda = \frac{3 \times 10}{300 \times 10} = 1 \times 10 \text{ centimeters}$$

Finding the wavelength of light at its lowest frequency requires solving this next equation:

$$\lambda = \frac{3 \times 10}{300 \times 10} = 0.1 \text{ centimeters}$$

The highest frequency produces the shortest wavelength (1×10 centimeters) and that the lowest frequency produces the longer wavelength. This is true for any frequency of electromagnetic radiation.

Usually, wavelengths that are as short or small as 1×10^{-7} (0.0000001) centimeters are expressed in other units called angstroms, designated by the symbol A. One angstrom equals 1×10^{-9} centimeters, therefore the smallest wavelength of light equals 10 angstroms.

Units of Measurement

Two methods or systems are used in the measurement of light. The first is the photometric system, concerned with the measurement of electromagnetic radiation known as visible light. However, as Fig. 6-1 illustrated, the complete spectrum of light contains a considerably wider range of frequencies than that of only the visible portion of frequencies. The radiometric system is used when considering the entire light spectrum. It includes all frequencies of light from infrared to ultraviolet.

As you can see from Fig. 6-1, the visible part of the electromagnetic spectrum that we call light is only a very small percentage of the entire spectrum of frequencies that consist of light. The radiometric system of light measurement is usually the more popular method of measurement used since many optoelectronic devices operate not only with visible light but also above or below that portion of the visible frequency spectrum. A typical device that operates above the visible portion of the light spectrum is the photodetector. This optoelectronic device responds to ultraviolet light waves which are invisible and is therefore very useful in burglar alarms and night time security surveillance systems.

Manufacturers have taken some liberty in the use of terminology used in both the photometric and radiometric systems of measurement. Many terms used in the photometric system of light measurement are sometimes used improperly when dealing

Table 6-1.

Convert from	Convert to	Multiplication factor
candle/cm²	lamberts	3.142
lumen	spherical candle power	0.07958
footcandle	lumen/m²	10.764

with the radiometric system of light measurement. In such cases it's helpful to have a conversion table on hand like the one shown in Table 6-1. Initially, you can see that a knowledge of terms used in both systems is needed. This table can be used to convert from the photometric system to the radiometric system. Be cautious, however, when using a conversion table like this one because the radiometric spectrum or range contains all of the frequencies of light, whereas the photometric system is limited to a much narrower range of frequencies. Therefore, conversion from one system to another does not always yield the proper results.

Terminology can be rather confusing too, but in actual practice only a few terms are needed when viewing a manufacturer's specification sheet to determine the proper device for a specific circuit need. So that you can be familiar with more than just one or two terms, the following definitions give you an introduction to some of these terms and their definitions.

Luminous Power. Also referred to as luminous flux and represented by the symbol Φ_v, it is the amount of luminous energy produced by a source of light over a unit of time.

Lumen. The basic unit of measurement for luminous flux.

Illumination. The amount of luminous power striking a surface per unit area and represented by the symbol E_v. Also called illuminance.

Lux. The unit of measurement of illuminance. One lux equals one lumen per square meter.

Footcandle. A unit of measurement for illuminance, as is the lux, and is equal to one lumen per square foot, or 10.76 lux.

Luminous Intensity. The luminous flux per unit solid angle traveling in a specific direction from a light source and represented by the symbol I_v.

Luminous Exitance. Measured in lumens per square meter, it is the luminous flux that is emitted from a unit area of a surface and is represented by the symbol M_v.

Luminance. Also referred to as brightness, it is represented by the symbol I_L.

Up to this point the term light has been used in a very general way, but specific sources of light must be considered when calculating certain quantities and in the conversion of those quantities and units from radiometric to photometric and vice versa.

One type of light is called point source light and radiates from a fixed point outward in all directions from that point. This type of light source brings with it a set of formulas called point source formulas. The criteria for point source calculations is a function of the distance between the point source of light and the photo sensitive device or detector that senses that light. It must be equal to or greater than 10 times the diameter of the point source of light or 10 times the diameter of photodetector, whichever is greater. If this criteria is not met, then the applicable table to use is called the area-source light table and differs considerably from the previous table. The most obvious difference may be that units of measurement in the radiometric system are given in watts, perhaps a more familiar term to you, whereas units of measurement in the photometric system are given in candles and lumens.

LIGHT SENSITIVE DEVICES - PHOTODETECTORS

Light sensitive devices can be categorized as either photovoltaic or photoconductive. The photoconductive type is also referred to as a photoconductive cell because there is usually a single element within the device that performs its function in controlling current. Also, it requires an external biasing circuit. The photovoltaic device does not require an external biasing circuit and is actually capable of producing electrical energy from a light source.

The Photovoltaic Cell

The photovoltaic cell is a quantum detector and converts light energy directly into electrical energy. The amount of electrical energy produced is directly proportional to the intensity of the light; that is, the amount of photons striking the light sensitive material, hence the term quantum detector. You may already be familiar with this type of device because of its more popular name, the solar cell.

A basic operating principle of the photovoltaic cell is the fact that in the infrared region of the electromagnetic spectrum, heat is a form of energy. In effect, in an ordinary PN junction diode an increase in temperature results in an increase in electron-hole pairs. Light in the visible area of the electromagnetic spectrum also generates electron-hole pairs and does so when striking an exposed PN junction of a diode. The construction of a photovoltaic cell is shown in Fig. 6-2. It is a diode in which leads are attached to metal support areas. In addition, silicon is usually used as the semiconductor material.

Photovoltaic cells generate a voltage depending upon the wavelength and energy content of the light striking the exposed surface. This means that they are sensitive to certain colors and this sensitivity determines the depth into which photons are absorbed into the semiconductor material. If a photon has a high enough energy content, it adds energy into the atom in which it comes into contact. This added energy causes the atom to give up an electron in its valence shell. The electron is sometimes said to be knocked out of its shell and becomes a free electron. You may see here that what is actually happening is that the valence shell surrounding the atom now contains a hole and the atom becomes positively charged. An electron-hole pair is formed by the photon entering the exposed surface of the semiconductor material. This process is repeated again and again throughout the material. Not all of the photons entering the semiconductor material create electron-hole pairs. In fact, between photons not striking atoms and electron-hole pairs recombining, the typical solar cell is only about 10 percent

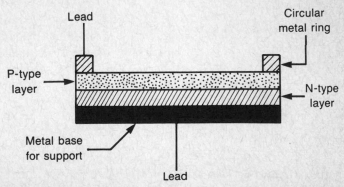

Fig. 6-2. A side view (cross-sectional) of the construction of a basic photovoltaic cell.

efficient in converting light energy into electrical energy. However, this conversion of light energy into electrical energy does indeed take place.

Those free electrons and holes created outside of the depletion region surrounding the PN junction are drawn into that region. Holes then migrate from the N-type material while electrons move in the opposite direction, from the P-type material to the N-type material. This movement of holes and electrons is actually current flow. Remember, current flow in a semiconductor material consists of both electron flow and hole flow. In effect, a small voltage is therefore produced across the PN junction of the diode. If a load is connected to this diode, current flows from the N-type material, through the load, and back to the P-type material. In effect, a small source of potential energy exists at the N- and P-type terminals of the photovoltaic cell just as in a battery's negative and positive terminals. In addition to silicon, other materials such as cadmium selenide (CdSe) are commonly used in the construction of the photovoltaic cell.

Typically, a photovoltaic cell needs about 2000 footcandles of light illumination to produce about 0.5 volts of electrical energy under a no-load condition. A number of these cells must be connected together in order to produce the required voltage and current for certain applications. Figure 6-3 shows the symbol used for the photovoltaic cell.

The Photoconductive Cell

The photoconductive cell, sometimes called a bulk photo-conductor or photoresistor, varies its internal resistance as the intensity of the light source striking it changes. Usually these cells have a negative coefficient, meaning that as the intensity of light increases, the internal resistance of the device decreases, allowing more current to flow through it. The intensity of the light controls the internal resistance of the photoconductive cell and thus, current flowing through it from an outside voltage source. There is, in practical terms, a nonlinear proportion between light intensity

Fig. 6-3. The symbol for a photovoltaic cell.

120

and internal resistance. Figure 6-4 illustrates the construction of a photoconductive cell. The semiconductor materials usually used in the manufacture of this type of device are cadmium sulfide (CdS) or cadmium selenide (CdSe), also used in the manufacture of some photovoltaic cells.

Here a very thin layer of the light sensitive material is applied to an insulating substrate made from a ceramic material. On top of the light-sensitive material, a layer of metal is applied, but a space is left down the center of this top layer in the form of a series of S shapes. Leads are then attached to each half of the top metal layer so that the device may be used in an electronic circuit. Then the cell is mounted into a metal or plastic case with a top glass window to allow light to enter. Figure 6-5 shows a typical photoconductive cell.

Only a very small change in illumination is needed to produce a very large change in the photoconductive cell's internal resistance. The device may exhibit 500 megohms or more with no illumination and only a few hundred ohms when the illumination is greater than 10 footcandles. Although this makes the photoconductive cell one of the most sensitive devices, it also has the disadvantage of persistency. This means that its resistance takes time to change with changes in illumination, or more technically termed, the device displays a history or memory effect. However, typical power dissipation of these devices is in a range of from 3 milliwatts to 300 milliwatts with voltage ratings in the range of from several millivolts to 300 volts dc.

Two of the most commonly used symbols for the photoconductive cell are shown in Fig. 6-6. This is simply a resistor with a circle around it and either visible light arrows pointing to-

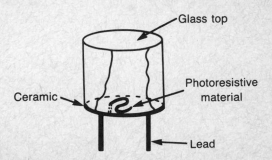

Fig. 6-4. A simplified drawing of the construction of a photoconductive cell.

Fig. 6-5. A typical photoconductive cell showing the photoconductive material as one continuous line from top to bottom (John Sedor Photography).

wards the circle or the Greek letter lambda inside the circle. Since the resistor symbol is used, and current can flow in either direction through a resistor, the photoconductive cell is also bidirectional in nature, making it suitable for use in either ac or dc circuits.

The Photodiode

The two types of light sensitive devices that have thus far been discussed have been the photovoltaic cell and the photoconductive cell. There is, however, a third category of devices that can exhibit the features of both of these devices called the photodiode. This device can be used as a photovoltaic cell, in which case it is said to be operating in its photovoltaic mode, or it can be used as a photoconductive cell, in which case it is said to be operating in its photoconductive mode. Figure 6-7 shows the symbol used to designate the photodiode. Although its construction is similar to the photovoltaic cell, its principle use is in the photoconductive mode of operation. Once again, the higher the frequency of the light and the greater its intensity, the less the internal resistance of the

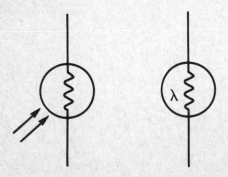

Fig. 6-6. Two of the more commonly used symbols for the photoconductive cell.

122

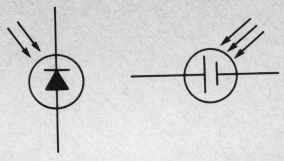

Fig. 6-7. The symbols used in electronic circuits for the photodiode.

device and therefore the greater the diode current. Sometimes these devices are called optical detectors and are a small part of the light receiving end used in fiber optic systems.

One of the more common types of photodiodes uses the PIN construction shown in Fig. 6-8. In this device an intrinsic layer of semiconductor material is placed between P-type and N-type semiconductors. The operation of this diode in the photoconductive mode is to reverse bias its depletion region which is considerably wider in this type of diode than in an ordinary PN junction diode. The extra width of this region makes it suitable for use at very low frequencies of light. Photons at these frequencies tend to penetrate the diode much deeper than at the higher frequencies and have less energy content also. Because of the much wider depletion region, there is a much better chance that these low energy photons produce the needed electron-hole pairs. A second advantage of the wider depletion region is that it provides for a more linear change in reverse current for a given change in light intensity and provides

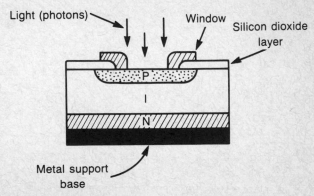

Fig. 6-8. Construction of a photodiode using the PIN method.

for a faster response to these changes in light intensity (less memory effect). This reduced memory effect is a result of lower internal capacitance of the device, again due to a larger depletion region.

Another method of construction of the photodiode is shown in Fig. 6-9. This process begins by forming an N-type layer onto a metal base. A silicon dioxide layer is placed on top of the N layer with a hole left in the center of the silicon dioxide layer through which a P-type region is diffused. Next, a metal ring is formed over the layer of silicon dioxide. The P-type region and the metal ring make electrical contact. Leads are attached to this window and to the metal base so that the device is usable in an electronic circuit. In the photoconductive mode this diode is operated in the reverse biased manner, shown in Fig. 6-9.

Biasing in this way creates a wide depletion region on either side of the PN junction. Photons entering this depletion region create electron-hole pairs just as they do in the photovoltaic cell. Because the diode is reverse biased and the charges on each side of the junction affect the electron-hole pairs (the voltage polarities of both the depletion region and the reverse bias voltage are in the same direction), the holes are attracted through the P region to the negative side of the reverse bias voltage source while the electrons are attracted through the N region of the diode to the positive side of the reverse bias voltage source. These electrons and holes now support a small reverse current through the photodiode. Additional photons (greater illumination) produce additional electron-hole pairs and therefore an increase in reverse current through the photodiode.

A number of characteristics are inherent in photodiodes. Remember, as the depletion region is widened, the capacitance between the N and P layers within a PIN photodiode is reduced, thus reducing the transit time of the electrons and holes to reach the ends of the device. This means that a PIN diode is much faster

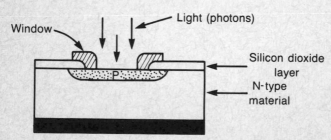

Fig. 6-9. Construction of a photodiode using the PN method.

than an ordinary PN junction diode. Some of these characteristics are better defined using terms that you should be familiar with. One of these terms is quantum efficiency, which expresses the performance of a photodiode, and can be expressed with the following equation:

$$\text{Quantum Efficiency} = \frac{\text{electrons}}{\text{photons}}$$

In an ideal situation, a perfect photodiode produces one electron for each photon entering the device and therefore the ideal quantum efficiency would be equal to 1. Of course, this simply is not the case, and the quantum efficiency is always less than unity. In fact the quantum efficiency does vary as the wavelength of the light source varies. At a wavelength of 4000 A the quantum efficiency of most PIN photodiodes is approximately equal to 0.3 and rises gradually to about 0.8 at a wavelength of about 8000 A. From there, the quantum efficiency drops off rather sharply up to 10,000 A where the quantum efficiency is close to 0.2.

Another measure of a photodiodes performance as a photodetector is called the noise equivalent power or NEP. This is the power necessary to produce a signal-to-noise ratio of unity with a noise bandwidth of 1 Hz. It is a measure of the minimum detectable signal level and is stated as follows:

$$\text{NEP} = \frac{\text{noise current } (A/Hz^{1/2})}{\text{current responsivity } (A/W)}$$

The lower the NEP, the lower the detection limit. As an example, if the input to the detector were 10^{-13} W, an NEP of 10^{-13} W/Hz$^{1/2}$ means that the output from the detector would barely be discernible from any random noise being produced by the detector itself.

Finally, performance can be measured by defining the responsivity of the photodiode. The responsivity is a measure of how much reverse current output is available for a given amount of input light energy.

$$\text{Responsivity} = \frac{\mu A}{mW/cm^2}$$

μA = output current referred to as photocurrent measured in microamperes.

mW/cm = irradiance or input radiant energy measured in milliwatts per square centimeter.

Once again, just as with quantum efficiency, responsivity is also a function of the wavelength of the light source. When the radiant power input is reduced to zero, the photocurrent of the photodiode does not also reduce to zero, only very nearly so. This is because of an extremely low leakage current that is characteristic of any PN junction diode. This low leakage is referred to as the photodiode's dark current and is usually measured in the nanoampere range.

Phototransistors

The phototransistor is a device that has the ability not only to detect light like the photodetector, but also the ability to provide amplification as well, all within a single device. It is constructed in a manner similar to any transistor and used like a photodiode. Construction of the typical phototransistor is shown in Fig. 6-10. Basically an N-type substrate serves as the phototransistor's collector. The base is formed by diffusing a P-type region into the substrate. Finally, the emitter is formed by diffusing an N-type material into the P-type base region. Three leads are then connected, to each area of the device. As you can see, construction is similar to that of an ordinary NPN transistor. The exception is that a

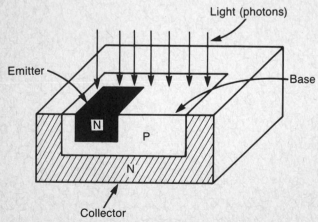

Fig. 6-10. Construction of a typical phototransistor is shown here.

transparent window is mounted over the device so that light can strike the surface of the device.

Operation of the device is straightforward. Photons striking the surface of the device generate electron-hole pairs in the depletion region of the collector-base junction. Since the collector-base junction is reverse biased, the holes are drawn to the base area and the electrons are drawn towards the collector. Since the emitter-base junction is forward biased, holes flow from the base to the emitter while electrons flow from the emitter to the base. As you can see, this is typical transistor action. The emitter injected electrons are drawn across the base region to the more positive collector region. In other words, this collector current is light-induced. Also, the photons entering the device causing electron-pairs add to the base current. If this particular phototransistor is now connected in the common emitter configuration, the device's collector current can be calculated by multiplying transistor's β by the light induced base current.

The electrical characteristics of the phototransistor are shown in Fig. 6-11. This is a graph for several different values of irradiance as stated in mW/cm^2.

The phototransistor is used very much like the photodiode in controlling current. The phototransistor, however, can handle far greater levels of current than the photodiode. It is also more sensitive giving it a wider range of applications. However, it does not respond as fast to changes in light levels as the photodiode does. Figure 6-12 shows the symbol used to designate the phototransistor and a circuit drawing showing a properly biased phototransistor.

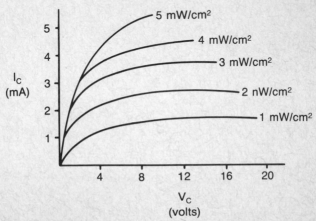

Fig. 6-11. The electrical characteristics of a typical phototransistor.

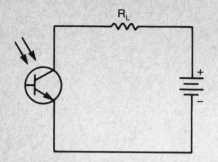

Fig. 6-12. The symbol used for the phototransistor and how it is biased in a circuit.

LIGHT EMITTING DEVICES

There are a number of light sources available today that can be used in conjunction with photodetectors or used by themselves, typically as light displays, in electronic circuits. Light sources include, but certainly are not limited to, such devices as incandescent light (the light you use at home that radiates from a typical light bulb), natural light from the sun, light emitting diodes, and liquid crystal displays that are now used in watches and other displays in preference to the older style light emitting diode (LED) display.

Probably the most popular form of device used specifically for monitoring the status of an electronic circuit or piece of electronic equipment is the LED. You can tell at a glance what the system is doing and if it is performing properly simply by viewing a series of LEDs and checking to see if they are on, off, or even pulsating at a specified rate or frequency. The greatest advantage in using the LED is that it is a solid state device and capable of a long operating life.

Light Emitting Diodes

When studying photodetectors you learned that light energy can be converted into electrical energy when photons entering a semiconductor material cause the formation of electron-hole pairs. In a similar series of events, a PN junction diode can emit light when these electron-hole pairs are recombined. Light emission, in this case, is referred to as electroluminescence. When a free electron recombines with an atom, the energy level of the electron falls from a high content or state to a lower state. This change in energy level releases a photon with a wavelength that directly corresponds to the energy level difference due to this transition. In the operation of the LED these higher energy electrons are

128

provided by forward biasing the diode. This causes electrons to be injected into the N region of the diode and holes to be injected into the P region of the diode, as illustrated in Fig. 6-13.

The electrons and holes injected into the diode from the outside forward bias voltage source combine with the majority carriers near the junction of the diode. Since the amount of material between the top surface and the junction is minimal compared to other sections of the diode, it is here that most of the external light is observed when the recombination radiation is emitted in all directions. The wavelength of the emitted photons is also a function of the semiconductor material used. And of course, as you may have guessed, you can see the emitted light from an LED but not from an ordinary PN junction diode because LEDs are made from semitransparent semiconductor material rather than opaque silicon.

LEDs can exhibit a number of different colors. The most popular colors are red, green, and amber, or yellow. Typically, color is a function of wavelength. However, the wavelength can be varied by varying the dopant in the LED. As an example most LEDs are made today using gallium phosphide (GaP) and gallium arsenide phosphide (GaAsP). Various concentrations of these materials produce varying wavelengths and corresponding colors. Increasing the percentage of GaAsP in an LED reduces the emission wavelength proportionately, yielding a different color. Infrared LEDs are best made from GaAs because the emission wavelength of the GaAs

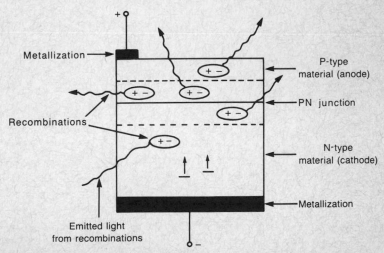

Fig. 6-13. Operation and cross-sectional view of a basic light emitting diode (LED).

LED is very nearly the same as the spectral response of a silicon photodetector, making these two devices compatible in security systems. The GaAs LED emits an invisible infrared light easily detected by a silicon photodetector. Anyone stepping in the path of this infrared light interrupts the light and, depending on the circuitry used, may set off a loud alarm. Since this is infrared and therefore invisible, it is ideal for nighttime security systems.

Figure 6-14 shows the basic construction of the LED. GaAs is used as the substrate. Next an epitaxial layer of GaAsP is grown and doped with the desired level of GaP. During this epitaxial growing period, an N-type impurity is then added to make this second layer an N-type material. An insulating layer is placed on top of this second layer. A window is then etched into the center of this insulating layer. Finally, a PN junction is formed by diffusing a P-type impurity through the window into the epitaxial layer. To allow connection to the LED leads are attached to the P-type region through an electrical contact and to the bottom of the substrate. Now the LED must be mounted in a suitable enclosure so that the emission of light from the LED is optimized. The package must contain a lens system because the quantity of light emitted is very small indeed. Figure 6-15 shows the basic construction of the packaging used for most standard LEDs.

There is an anode lead and a cathode lead and in this sense this diode must be forward biased for proper operation. The LED chip itself is very small. Some lenses are of the diffused type which helps to diffuse the light rather than project or optimize it in a particular direction. Other types of lenses help to direct the LED emission in a direction suitable for certain applications where the LED is being viewed from a specific angle.

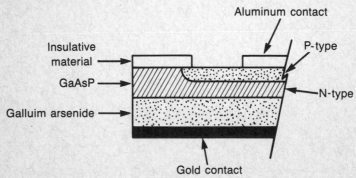

Fig. 6-14. Basic construction of a GaAsP LED device.

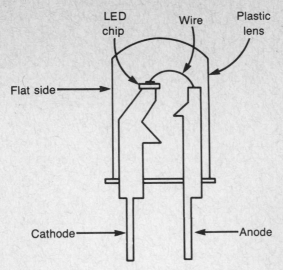

Fig. 6-15. Basic mounting package for an LED.

Mounting the package requires pushing the assembly through the right size hole or placing the LED in a bezel and snapping it into place. Figure 6-16 is a photo of several different LEDs, including one with a mounting bezel intact.

The V-I characteristic curve of a typical LED is shown in Fig. 6-17. There is a somewhat sharp knee in the area of around 1.2 volts. Then the diode begins to conduct current while the voltage across the device remains constant at about 1.6 volts. The greater

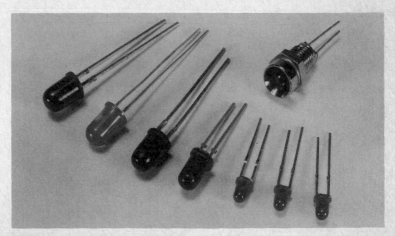

Fig. 6-16. Some typical LED packages, including a bezel mount for mounting an LED into a panel (John Sedor Photography).

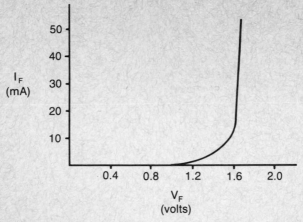

Fig. 6-17. V-I curve for a typical GaAsP LED.

the current through the LED, the greater the relative radiant power output, or brightness, of the device. However, precautions must be taken so that excessive current is not allowed through the LED. Just as with any diode, the manufacturer's specifications must not be exceeded or the device is damaged.

An LED with its standard symbol is shown in Fig. 6-18. Remember, too much current can damage the device. It is necessary to determine the size of the series resistor needed to limit the current through the LED to approximately 50 mA. This level of current is given because the relative radiant output power is almost 100 percent with 50 mA of current flowing through a typical LED. To calculate R_S, where R_S is greater than 40 ohms, the following equation is used:

$$R_S = \frac{V - 1.6}{I_F} - 5 \ \Omega$$

Fig. 6-18. The symbol used for the light emitting diode (LED).

R_S = Series limiting resistor
V = Applied bias voltage
I_F = Forward current through the LED
1.6 = Typical voltage drop across an LED
5Ω = Typical internal resistance of LED

Solving then for R_S:

$$R_S = \frac{12 - 1.6}{0.05} - 5$$

$$R_S = 208 - 5 = 203 \text{ ohms}$$

An interesting aspect to LEDs is that most of the photons that are emitted from the diode never actually leave the device itself. Most are absorbed by the semitransparent material and many are actually reflected back into the semiconductor material. This is because a photon must leave the material at an angle of 17° or less, called the critical angle. Any angle greater than 17° causes the photon striking the surface to be reflected back into the LED material. This critical angle can almost be doubled by mounting the LED chip into an epoxy or plastic package.

LED Displays

One of the more popular methods of use of the LED is in the LED display. Figure 6-19 shows several LED displays referred to as seven-segment displays, because the display is separated or made up of seven sections, each of which is capable of emitting light. Depending on which segment is lit, the LED displays all whole numbers from zero to nine. The construction of one of these LED displays requires the use of seven LED chips placed inside the package. Mounted over each chip is a section of internal package called a light pipe which reflects the emission from the LED chip up towards the top surface of the display. As you can see from the figure, some displays also carry decimal points that can be used to signify a whole number, tenth of a number, or one hundredth of a whole number.

Another popular use of the LED is in a solid state device called an optical coupler, a small integrated circuit package (shown in Fig. 6-20) that contains an infrared LED and a phototransistor. These

Fig. 6-19. Some typical LED displays which display numbers, letters, and a bar graph (John Sedor Photography).

two devices are separated by a special light transmitting glass. The electrical connection between the devices is through the infrared light transmitted from the LED to the phototransistor. When the input signal to the LED varies, the infrared emission from that device also varies in intensity. This varying light is detected by the phototransistor and its conduction therefore varies as the intensity of the infrared light striking its surface. In this way a signal can pass from one circuit to another with a significant degree of isolation between the two.

Liquid Crystal Displays

Recently, a new method of displaying information has come about that is significantly different from the LED display previously discussed. The advantages of the liquid crystal display as compared to the LED is its lower power requirements and its easier viewing in high ambient lighting. It is also usually lower in cost than a comparable LED and maintains excellent contrast. On the negative side, it has poor visibility in low ambient lighting and usually requires an ac drive rather than dc as does the LED.

There are two different types of LCDs: those using dynamic scattering to obtain a display, and those using a field effect to obtain

Fig. 6-20. These integrated circuit packages contain an infrared LED and a phototransistor each and are called optical couplers or optical isolators (John Sedor Photography).

Fig. 6-21. This LCD is found in watches and in many of today's applications including this solid-state digital multimeter (John Sedor Photography).

a display. In effect, LCDs do not emit light, but rather rearrange externally generated light to produce an image. In the field effect LCD, a liquid crystal fluid is held between two glass plates with the segments located on the top plate. When an electrostatic field is applied to the LCD contacts, the liquid crystal fluid is polarized and, through molecular action of the fluid, dark digits appear on a light background. A liquid crystal display is shown in Fig. 6-21. This is the type of display found in most applications today.

The other type of liquid crystal display is the dynamic scattering type. In this type of LCD, the liquid crystal fluid is normally clear. When an electric field is applied, the external light striking the LCD surface is scattered due to turbulence from ion activity of the fluid rendering a white display on a dark background. This type of LCD is not seen very often.

The number of uses for LEDs and LCDs is endless. From security systems and circuit isolation to detecting the presence of an object or displaying information, LEDs and LCDs are only beginning to reveal their potential in the optoelectronic area of solid state electronics.

Chapter 7

Basic Bipolar Transistors

Bipolar transistors were covered in Chapter 4, with some theory of beta and alpha gains for particular transistor configurations. There is a great deal more to transistors as amplifiers, and you will need to know more about the three basic circuit configurations, proper biasing of these transistor configurations, how transistors are coupled when more than one is used in a circuit (as is almost always the case), and the different classes in which amplifiers are operated. All of this information helps you better understand the bipolar transistor as an amplifier and how to design, test, and troubleshoot amplifier circuitry. This chapter deals with these aspects of bipolar transistor amplifiers. The following chapter deals with specific types of amplifiers such as audio, video, and rf amplifiers. First, a basic understanding of bipolar transistor amplifiers is in order.

THE IMPORTANCE OF AMPLIFIERS

It is very difficult to detect the music from the tone arm needle of a record player without the addition of an amplifier to properly reproduce the sound and amplify it at the same time. The same is true when speaking into a microphone. The speakers would be silent if it were not for the power amplifiers located between the microphone and the speakers. Amplifiers are used in all types of electronics equipment that include, but certainly are not limited to, military, industrial, and commercial applications.

136

Definition of an Amplifier

An amplifier is actually not a single solid state device, but a combination of components with a transistor as the central device that reproduces and increases the level of an electronic signal. An amplifier does not really produce a greater power or energy at its output from what is present at its input. It amplifies by allowing a very small input signal to control a much larger outside source of energy, usually connected to the output of the device. An amplifier does not create a large output signal for a very small input signal.

Amplifiers are designed to faithfully reproduce the input signal, but at a much larger level and with no irregularities. That means the output signal should look exactly like the input signal, but it is doubtful that any amplifier can amplify and reproduce an input signal without some distortion. In most cases distortion can be minimized. Some amplifiers actually distort the input signal intentionally. They may be designed to produce a square waveshape output from a sine wave input, or even a triangular waveshape output from a sine wave input. These types of amplifiers are actually generators because they generate a specific output for a particular signal input.

An amplifier is a combination of components that when properly connected together, amplify an input signal to a much larger output level. Some of the other components that make up the amplifier are capacitors, resistors, and in some cases, inductors. How these work together to produce amplification is the subject of this chapter.

Where Are Amplifiers Used?

Amplifiers are used in almost every piece of electronics equipment that you may see or use. Not all amplifiers are made up of transistors and other discrete components. Some amplifiers are packaged in integrated circuit form, but they are still found in many different types of electronic equipment, including television sets, radios, stereo systems, tape recorders, and electronic keyboard instruments. In televisions, there are several different types of amplifiers: video amplifiers to amplify the picture level before it is sent to the picture tube for display, rf amplifiers to amplify the signal coming from the cable company or the local TV station, and audio amplifiers to amplify the sound signal in the TV broadcast signal prior to it being sent to the loudspeaker. In tape players and recorders, the playback heads pick up the very small magnetic signals from the tape. It is these very small signals that must be amplified from the playback heads so that you can hear your favorite music.

Probably a more important need for amplification has come in the way of two-way radio communications. Speaking into a radio microphone requires that the input voice signal be amplified many times by audio amplifiers and then mixed with high frequency signals and transmitted via a powerful rf power amplifier to a receiver somewhere, usually many miles away. Ham radio operators attest to the importance of high power and even low power amplifiers in obtaining contact with individuals sometimes thousands of miles apart. In telecommunications, the cellular car telephone has become increasingly popular among our ever-mobile society. It too is simply a two-way communications device requiring amplification for proper operation.

You can probably think of a hundred or more ways in which amplifiers are utilized, but the important point to remember here is that amplifiers are probably the single most common circuit found in electronic circuits today.

AMPLIFIER CIRCUIT CONFIGURATIONS

There are three basic amplifier circuit configurations: the common-emitter, common-base, and common-collector or emitter follower. These circuits were discussed in Chapter 4 but this section deals with the characteristics of each of these circuit configurations beginning with the common-emitter circuit. These characteristics include dc and ac beta, voltage gain, power gain and their equations, and input and output resistances.

Keep in mind that each amplifier, no matter how it may be configured, must have two input leads and two output leads. Since a transistor has only three leads and four have just been stated as a requirement, then one of the leads is common to both the input and output signal paths, hence the terms common-emitter, common-base, and common-collector. Also, since the common lead is usually connected to ground, these circuit configurations are also called grounded-emitter, grounded-base, and grounded-collector. Ground in this case refers to ac ground, not necessarily dc ground. As far as an ac signal is concerned, a lead does not have to be connected directly to circuit ground to be considered grounded. A lead can still be at an ac ground potential even if first connected through a capacitor bypassing a battery because the capacitor passes ac directly to ground. In the circuits to follow, the ground that the lead is connected to is referred to as ac ground unless connected directly to circuit ground. In that case the lead is considered to be at both ac and dc ground potential. This is an important concept and should not be overlooked.

Common-Emitter

Figure 7-1 is an example of how an NPN transistor is connected in the common-emitter configuration. The emitter is connected directly to ground. The input signal is connected between emitter and base, and the output is developed across emitter and collector. The emitter in this circuit can be seen as the common or grounded lead.

In this particular circuit drawing, the junction biasing voltage sources are shown as separate supplies. Later on, when different types of biasing schemes are discussed, the battery symbols are eliminated, as you will almost always find in electronic circuit design. For now, this type of drawing makes it easier to see what actually is happening with the amplifier circuitry.

Referring to Fig. 7-1, notice that the collector-base junction is reverse biased while the emitter-base junction is forward biased. This is the normal way in which these junctions are biased in order for the transistor to operate properly. Base bias supply V_{BB} is the voltage source used to forward bias the emitter-base junction while collector bias supply V_{CC} is used to reverse bias the collector-base junction. Applying an input signal causes V_{BB} to be connected, in effect, across the emitter-base junction of Q_1, because the signal source provides a path for current through its internal resistance. This current flows through the junction, the signal source, and V_{BB}

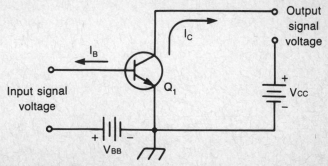

Fig. 7-1. An NPN transistor, properly biased, is shown here in the common-emitter circuit configuration.

and is referred to as base current, I_B.

It may not be apparent that VCC is reverse biasing the collector-base junction of Q_1. It is, however, because the emitter-base junction being forward biased presents only a minimal resistance between the negative side of VCC and the base of Q_1. Therefore, with an external load connected across the transistor's emitter and collector, current flows from the negative side of VCC, through the emitter-base junction, the collector, the load, and back to the positive side of VCC. Current flows from emitter to collector because the emitter-base junction is forward biased allowing a very small I_B to flow. The emitter current actually represents 100 percent of the current flowing through the transistor with about 5 percent going out of the base and the rest, about 95 percent, called collector current or I_C, going out of the collector. This can also be stated as follows:

$$I_E = I_B + I_C \quad \text{or} \quad I_C = I_E - I_B \quad \text{or} \quad I_B = I_E - I_C$$

The relationship between I_B and I_C indicates that a very small base current allows a very large collector current to flow. In essence, this circuit produces an increase in current and therefore exhibits current gain. Therefore it follows that an increase in I_B causes an increase in I_C, and a decrease in I_B causes a decrease in I_C. Decreasing I_B to zero just about cuts off current flow through the transistor except for a very small leakage current. Increasing I_B greatly only increases I_C to a level where eventually it begins to taper off. In other words, the transistor simply will not conduct any more current no matter how much I_B is increased; it is then said to be operating in saturation.

If a single value of I_C is measured and compared to a single value of I_B, with VCE held constant at the point of measurement, the transistor's steady state or dc current gain can be calculated simply by dividing I_C by I_B or:

$$\text{dc current gain} = \frac{I_C}{I_B}$$

The dc current gain here is called the transistor's dc beta. Ac beta of a transistor is a measure of the transistor's current gain when a ratio is determined after measuring a change in collector current for a corresponding change in base current. This means an input signal is varying in level and causing a corresponding change in collector current. This is actually the current gain most manufacturers specify and the one that is used most often in designing amplifier circuitry. It is found by the following equation:

$$ac\ beta = \frac{\Delta I_C}{\Delta I_B}$$

In many cases the dc beta and the ac beta are very nearly equal in value and for a single transistor can vary anywhere from 5 to 200 or more. Theoretically, beta values are always determined with VCE held constant under no load conditions in the common-emitter mode. In practice, a load resistor is used so that collector current flows, but the current gain is still a rather accurate indication of the transistor's amplifying factor. Also, when discussing dc beta, the symbol h_{FE} is used and h_{fe} is used when signifying ac current gain.

To obtain voltage gain, an output load resistor must be used so that I_C can flow through R_L and develop an output voltage drop. This is shown in Fig. 7-2. Notice that R_L is connected in series with VCC and that the output voltage, V_{out}, is taken across the transistor and actually represents VCE. Figure 7-3 is a graphic representation of current and voltage inputs and outputs and also the voltage drop across R_L. V_{out} is 180 degrees out of phase with V_{in}, characteristic of voltage gain in the common-emitter circuit configuration.

Since VBB is connected to Q₁'s emitter-base junction through the input signal source it can be set to a value to cause a specific

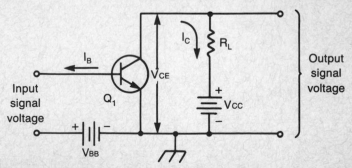

Fig. 7-2. A common-emitter circuit with a load resistor, R_L, used to develop an output voltage.

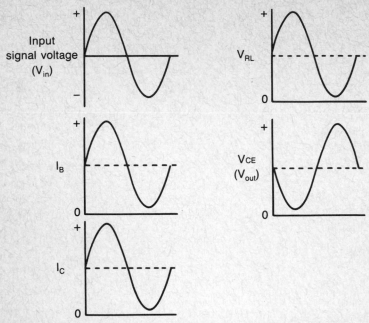

Fig. 7-3. The signals that are used to represent the voltages and currents in a common-emitter circuit where I_B, I_C, VRL, and VCE are no signal, steady state values.

I_B to flow when V_{in} is equal to zero. This specific value will then determine a specific value of I_C which will flow through R_L. Therefore, if V_{in} varies, VRL also varies. Now, R_L is connected in series with Q_1 and both are connected across VCC. That means VCC equals VRL plus VCE. If V_{in} varies, VRL varies and thus VCE (V_{out}) also varies. V_{out} is a reproduction and amplification of V_{in}, but it is out of phase. The out of phase output can easily be explained.

A positive input signal voltage aids the positive base bias voltage source, VBB. I_B therefore increases as the input signal increases. I_C also follows this increase since I_B determines I_C. Also, since I_C flows through R_L, VRL also increases with an increase in the input signal voltage. Since VCC equals VRL plus VCE, VCE (V_{out}) must decrease to maintain VCC, otherwise VCC would have to vary. Therefore to account for a constant VCC when VRL increases, V_{out} decreases. If now the input signal voltage goes negative, just the opposite occurs with I_B, I_C, VRL and V_{out}. Therefore I_B, I_C, and VRL are in phase with the input signal while V_{out} remains 180 out of phase. All of this is shown in Fig. 7-3.

However, phase reversal does not negate voltage gain. Since V_{out} is much larger than V_{in}, this circuit does provide voltage gain, or A_v, and can be calculated as follows:

$$A_v = \frac{V_{out}}{V_{in}}$$

Incidentally, ac signal values are usually used in determining A_v. These may be peak-to-peak, peak, average, or effective values.

As you may have guessed, current gain and voltage gain also provide power gain and this is true for the common-emitter amplifier. In this configuration, power gain is found by the following equation:

$$A_P = \frac{P_{out}}{P_{in}}$$

A_P = power gain
P_{out} = output power
P_{in} = input power

Power gain, A_P, can also be found by multiplying the transistor's ac beta, β, by its voltage gain, A_v, or $A_P = \beta A_v$. Extremely high power gains in the neighborhood of 10,000 or higher are achievable with the common-emitter configuration simply because this arrangement provides both current and voltage gain.

In essence, the input resistance of the common-emitter transistor is relatively low because the emitter-base junction is forward biased, presenting a small resistance to input signal current. However, this input resistance is an ac quantity because it is a function of a changing input signal and therefore varies slightly as I_B changes. Also keep in mind that additional components are used in biasing a transistor and these add to or subtract from the transistor's total input resistance.

Since the transistor's collector-base junction is reverse biased, you would expect a high output resistance from the common-emitter circuit, and this is true. It is usually about 50 k ohms. Again, this output impedance represents an ac quantity because it is the opposition to a changing current and therefore, ac. Output resistance is also a function of load resistance since R_L varies the transistor's total output resistance and is also a function of V_{CE}.

The characteristics of the common-emitter circuit can be summarized as follows:

- High voltage gain
- High current gain
- High power gain
- Low input resistance
- High output resistance

Common-Base

Figure 7-4 shows a circuit diagram for the common-base circuit configuration. Once again, the battery symbols are used to show bias voltage sources so that you can better understand the operation of this configuration. In this circuit, the base lead is common to both input and output signals, with the emitter-base junction forward biased and the collector-base junction reverse biased as before. Here V_{EE} provides emitter-base junction forward bias and V_{CC} provides collector-base junction reverse bias. The following equation still holds true for this configuration: $I_E = I_B + I_C$. This means that a change in I_E causes a corresponding change in I_B and I_C.

In this circuit, I_E controls I_C. To allow a proportional change in I_C for changes in I_E, V_{EE} and V_{CC} are adjusted so that the transistor operates in its linear region. Current gain in the common-base circuit is found by dividing I_C by I_E and is referred to as the alpha of the transistor. Dc alpha can be calculated using fixed values of I_E and I_C while ac alpha can be calculated using changing values of I_E and I_C. Alpha is represented by the symbol (α).

Mathematically, if $I_E = I_B + I_C$, then I_C must be smaller than I_E because I_C alone does not equal I_E. Therefore the transistor's

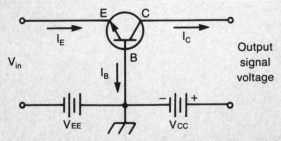

Fig. 7-4. A properly biased common-base circuit using an NPN transistor.

current gain is less than 1, or is said to be less than unity. A typical dc alpha might be 0.95 or 0.98. Therefore in the common-base circuit there is no current gain, however, the dc alpha is still referred to as gain and is represented by the symbol h_{FB}. The ac alpha gain is represented by the symbol h_{fb}.

Not having current gain might give you the idea that no voltage gain could then be found with this type of circuit; however, a substantial voltage gain as high as 1000 can be provided by the common-base circuit. This is accomplished by developing a high voltage drop across a high value load resistance through which I_C must flow. This voltage drop represents a part of the total output voltage and is shown in Fig. 7-5. Here, VRL is a part of VCC and also VCB.

To see how amplification takes place, assume a sinusoidal input waveform applied to the input of the circuit. You can follow what happens by looking at Fig. 7-5 and also at the waveforms of Fig. 7-6. The positive alternation of the input signal opposes VEE causing I_E to decrease in value. This causes I_C to decrease in value since I_C is controlled by I_E. Since I_C decreases in value and I_C flows through R_L, VRL will also decrease in value. At any given point in time VCC equals VCB plus VRL. Therefore, if VRL decreases, VCB must increase in value. VCB actually represents V_{out} and therefore it can be stated that an increase in the input signal voltage causes a corresponding increase in the output voltage. There is, in this case, no phase inversion, but definitely voltage gain. And of course, just the opposite occurs as the input signal voltage goes from positive to negative.

Since there is less than unity gain as far as current gain is concerned and voltage gain can be quite high, there is actually a

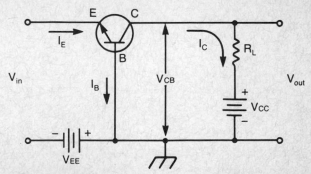

Fig. 7-5. A properly biased NPN transistor in the common-base configuration with an output load resistor.

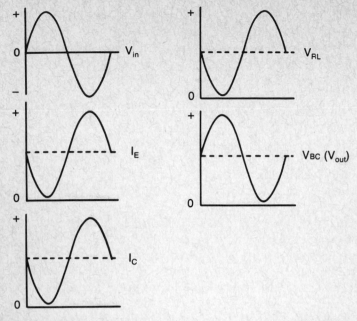

Fig. 7-6. These are the signals that are used to represent the voltages and currents in a common-base circuit where I_E, I_C, V_{RL}, and Vout are no signal, steady state values.

moderate power gain for the common-base circuit. Power gain can be found with the following equation:

$$A_P = A_v$$

A voltage gain of 1000 and a current gain of 0.98 yield a power gain of 980.

Remember that I_E is the total current flowing through the transistor meaning that the emitter-base junction offers a very small input resistance to the input signal across this junction and varies according to various other components connected at the input for different biasing arrangements. It can be as low as a few ohms.

Output resistance, on the other hand, is very high in this circuit configuration simply because the collector-base junction is reverse biased and can be higher or lower depending on other components in the output circuit. However, it can be as high as 1 megohm.

Summarizing these characteristics, the common-base circuit offers:

- Low current gain (less than unity)
- High voltage gain
- Moderate power gain
- Low input resistance (lower than C-E)
- High output resistance (higher than C-E)

Common-Collector

Figure 7-7 shows the circuit configuration for a common-collector circuit, again with bias supplies shown using battery symbols. In this circuit the input current is represented by the base current I_B, while output current is represented by emitter current, I_E. Varying I_B varies or controls I_E and in effect I_C. In the common-collector circuit, I_E flows through the load and the output voltage is developed across a load connected between the emitter of the transistor and circuit ground, shown in Fig. 7-8.

Current gain in this configuration is substantial since a very small I_B can control a very large I_E. In fact, it is higher than in the common-emitter configuration and is found by adding 1 to the beta of the transistor as if it were connected in the common-emitter mode: current gain $= 1 + \beta$.

Voltage gain is just slightly less than the input signal voltage. The output voltage follows the input voltage signal in phase and is almost the same amplitude and this configuration is therefore called an emitter follower. Voltage amplification does not occur because V_{RL} actually opposes any changes in V_{in}. When a sinusoidal input signal voltage is applied, the positive portion of

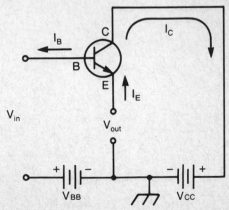

Fig. 7-7. A properly biased NPN transistor in the common-collector configuration.

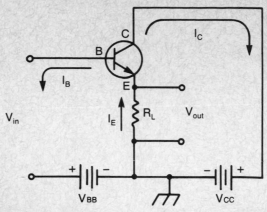

Fig. 7-8. A properly biased NPN transistor in the common-collector configuration with an output load resistor.

this signal causes the base to increase in its positive value with respect to ground. At the same time, the emitter also becomes more positive because it is in phase with the input signal. The following action continues to occur when the input signal is negative, but in the other direction. These simultaneous changes in input and output signals tend to oppose each other. The output voltage, rather than opposing the input voltage completely, remains at a level just slightly lower than the input signal voltage, allowing current to flow through R_L, developing an output voltage.

Figure 7-9 shows the relationship among V_{in}, I_B, I_E, and V_{RL}

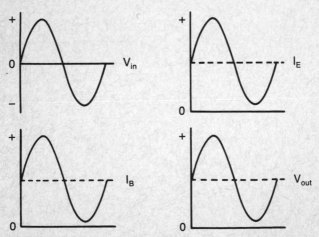

Fig. 7-9. The signals that are used to represent the voltages and currents in a common-collector configuration.

148

or V_{out}. V_{out} is so close to V_{in} in level that for all practical purposes the voltage gain of the common-collector circuit is unity or 1.

Once again, a moderate power gain can be realized because of the large current gain, but slight loss in voltage gain. Power gain can be found with the following equation:

$$A_P = (1 + \beta)A_v$$

Remember, Ohm's law states that power, P, is equal to voltage times current. Power gain then, is a function of voltage gain times current gain. Everything goes back to the basics of Ohm's law!

The input resistance is very high for the common-collector circuit because the input signal will see both R_L and the resistance of the emitter-base junction. Since R_L is considerably larger than the emitter-base junction resistance it actually represents the input resistance. However, the total input resistance is equal to the transistor's beta multiplied by R_L, or βR_L. This combination usually results in an input resistance of up to 500 k ohms.

The output resistance of this configuration is very low, around 500 ohms, and is a function of a number of factors including beta, R_L, and the intrinsic resistance of the input signal source.

The common-collector circuit makes a good buffer from one circuit to another. Its high input resistance prevents it from drawing an excessive amount of current from the input signal voltage source and its low output resistance allows a low resistance load to be connected to its output without loading down the common-collector circuit itself.

Summarizing the characteristics of the common-collector configuration, you'll find:

- Very high current gain
- Low voltage gain (slightly less than unity)
- Moderate power gain
- High input resistance
- Low output resistance

AMPLIFIER BIASING

In practice you'll probably never see two separate bias supplies as shown in the three common circuit configurations. Usually a single voltage supply is used and represented on circuit schematics, and it is usually designated as VCC. Proper operation depends on taking portions of this single VCC supply and using them to supply the necessary biases for the transistor. This is accomplished in a number of different ways. Usually you will find that it is simply a matter of voltage divider concepts in supplying proper biases to the transistor. Since the common-emitter circuit is the most popular, its biasing scheme is discussed first.

Base Biasing

Figure 7-10 is an illustration of biasing a common-emitter circuit using a single supply, VCC. Notice that R_L and the transistor are in series with each other and are, in reality, across VCC. Therefore VRL plus VCE is equal to VCC. R_B is in series with the base and controls the amount of I_B flowing out of the transistor and up to VCC. The emitter-base junction of the transistor and R_B are in series with each other and again, are across VCC. Since the emitter-base junction normally drops around 0.7 volts, the remainder, or most of VCC, is dropped across R_B. Also connecting the positive side of VCC to the collector and base of Q_1 through R_L and R_B properly biases the transistor. In essence, base current, I_B, is controlled by the size of R_B and the value of VCC. This type of biasing arrangement is termed base biasing.

The input, as can be seen, is applied between the base and ground, or emitter lead, and the output is taken across the transistor's collector and emitter leads. As I_B varies, so does I_C, causing R_L to drop a portion of VCC called VRL. The remainder is dropped across the transistor and varies as I_B varies. Although

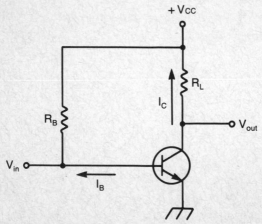

Fig. 7-10. A common-emitter circuit using a single supply and base biasing.

this type of biasing is simple to use and understand, it's not likely you'll see such a simple design because of its instability. As current through the device increases, so does temperature. If this temperature rise is not compensated for, the gain of the transistor is reduced and the transistor could conceivably burn up due to a degenerative process called thermal runaway. In effect, this circuit is said to exhibit thermal instability and needs other components to prevent this from happening.

Feedback Biasing

Feedback takes a portion of an output signal and feeds it back to the input to aid or oppose changes in the operation of the device. Opposing changes are referred to as negative or degenerative feedback. This kind of feedback is employed to produce thermal stability in an operating transistor and is called feedback biasing.

One method of supplying stability can be seen in Fig. 7-11. Here R_B is connected across the collector-base junction of Q_1 rather than from the base of the transistor to VCC. As temperature increases, so does the leakage current flowing from emitter to collector. This leakage current, called I_{CEO}, adds to I_C and this additional current causes VRL to increase. VCE decreases in value, and so do V_B and I_B. A decrease in I_B causes a decrease in I_C. In this way, I_C is brought back to within normal limits when temperature increases and I_C begins to increase. This is called collector feedback because the feedback signal is taken from the collector of the transistor.

A more stable circuit configuration is shown in Fig. 7-12. This is called emitter feedback since the feedback in this circuit comes

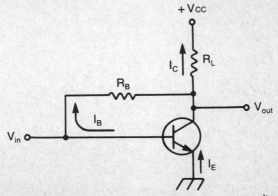

Fig. 7-11. A common-emitter circuit using collector feedback biasing.

151

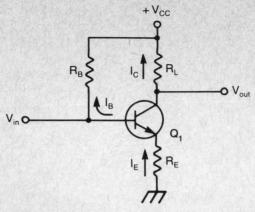

Fig. 7-12. A common-emitter circuit using emitter feedback biasing for temperature stabilization.

from the transistor's emitter. Once again, it is an improvement over the simpler method of collector feedback just presented. Examining Fig. 7-12 shows that R_B, the emitter-base junction of Q_1, and R_E are connected in series and are across VCC. For all practical purposes, the emitter-base junction can be ignored. When an increase in temperature causes I_C to increase, I_E increases as well. VRE then increases accordingly causing VRB to decrease. VRB decreases because VRE plus VRB plus the emitter-base junction voltage must all equal VCC. If one of them increases, the others

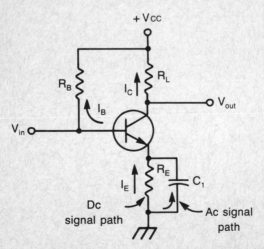

Fig. 7-13. A common-emitter circuit using emitter feedback biasing and an ac bypass capacitor for better stability.

must decrease. Since the emitter-base junction of 0.7 volts is being ignored, then when V_{RE} increases, V_{RB} must decrease. And, of course, I_B decreases reducing I_E and I_C back to within their normal limits.

In Fig. 7-13, a capacitor, C_1, has been added across R_E. C_1 is referred to as a bypass capacitor because it bypasses any unwanted ac signal component around R_E. Ac signal components are varying voltages that are prevented from being developed across R_E, R_L, or the transistor by using C_1 to shunt them around R_E. R_E is able to stabilize the transistor by continuously varying at a slow steady rate rather than at a rate dictated by an ac signal component.

Voltage Divider Biasing

Since stability is a key factor in amplifier circuit design, biasing by nature must be refined to a point where stability is achieved most effectively. Therefore a more effective design than emitter feedback biasing is shown in Fig. 7-14. This is called voltage divider biasing because R_1 and R_2 replace R_B and serve to act as a voltage divider for the base circuit of Q_1. V_{CC} is now equal to V_{R_1} plus V_{R_2} since they are in series with each other and are across V_{CC}. I_1 in this circuit is actually slightly larger than I_2 because it consists of I_B plus I_2. I_B is so small that it can really be ignored. R_1 and R_2 are designed so that I_1 and I_2 are considerably larger than I_B. In any event, the purpose of using voltage divider biasing

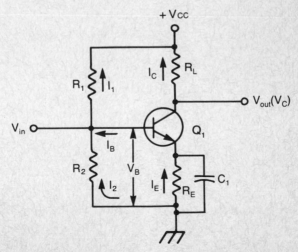

Fig. 7-14. Voltage divider biasing in a common-emitter circuit.

is to establish a stable dc bias point (stable dc level at the collector of the transistor) and a stable ac gain (voltage output at the collector of the transistor).

In the circuit of Fig. 7-14, the voltage across the emitter-base junction of Q_1 is equal to the difference between V_B and VRE. Usually V_B is slightly more positive than VRE. The junction voltage of Q_1 is usually constant at around 0.7 volts for silicon transistors and remains constant even with current changes through the junction. When an increase in temperature causes an increase in I_C and I_E, VRE increases causing the emitter to become more positive, reducing the emitter forward bias voltage. This causes I_B to decrease, reducing I_C and I_E back to their normal values.

To design a circuit like that of Fig. 7-14, there are a number of rules to follow:

- Use an R_1 value that is 9 times the value of R_2
- Use an R_2 value that is 10 times the value of R_E
- Use an R_L value that will drop 0.45 the value of VCC
- Use a value of R_E that will drop 0.1 the value of VCC
- Select an I_C that is within the manufacturer's specifications
- Select C_1 to have an X_C that is equal to 0.1 times the value of R_E at the lowest operating frequency. Use the formula:

$$C_1 = \frac{1600000}{fR_E}$$

As an example, assume the following values of in Fig. 7-14:

$$VCC = 15 \text{ volts}$$
$$I_C = 1 \text{ milliampere}$$
$$\text{Lowest frequency} = 40 \text{ hertz}$$

Therefore:

$$VRE = 0.1 \times VCC = 0.1 \times 15 \text{ V} = 1.5 \text{ volts}$$

$$R_E = \frac{1.5 \text{ V}}{0.001 \text{ A}} = 1500 \text{ ohms}$$

$$\text{V}_{\text{RL}} = 0.45 \times \text{V}_{\text{CC}} = 0.45 \times 15 \text{ V} = 6.75 \text{ volts}$$

$$R_L = \frac{6.75 \text{ V}}{0.001 \text{ A}} = 6750 \text{ ohms}$$

$$R_2 = 10 \times R_E = 10 \times 1500 = 1500 \text{ k ohms}$$

$$R_1 = 9 \times R_2 = 9 \times 15 \text{ k} = 135 \text{ k ohms}$$

$$C_1 = \frac{160000}{(40)(1500)} = 270 \ \mu\text{F} \times 0.1 = 27 \ \mu\text{F}$$

A note here about voltage gain, A_v. When C_1 is omitted from the circuit of Fig. 7-14, A_v is equal to R_L divided by R_E. However, when C_1 is in the circuit, A_v is equal to the transistor's beta, β, times the collector resistor R_L, divided by the emitter resistor R_E, or:

$$A_v = \beta \ \frac{R_L}{R_E}$$

AMPLIFIER COUPLING

In many cases a one-transistor amplifier simply does not provide enough gain to a particular circuit application. Two or more transistor amplifiers must be cascaded or coupled together to provide the needed amplification. Several methods are employed in coupling amplifiers together. Each has its own advantages and disadvantages, and each is used for specific purposes.

Direct Coupling

In order to amplify very low frequencies down to and including dc, direct coupling is employed. This means connecting the output of one transistor directly into the input of the other, as shown in Fig. 7-15. The collector of Q_1 is connected directly to the base of Q_2. R_4 performs the dual function of providing collector current for Q_1 and base current for Q_2 and is usually a resistor with a high value.

With an ac signal on the base of Q_1, the varying output of Q_1

155

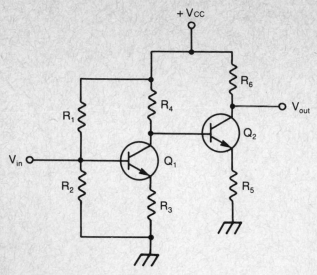

Fig. 7-15. Two stages of amplification using direct coupling.

is felt directly on the base of Q_2. The collector of Q_2 follows the input of Q_1, but since Q_1 phase inverts the input signal by $180°$, the output of Q_2 is in phase with the input to Q_1.

The biggest disadvantage to this type of coupling arrangement is that any temperature instability at the collector of Q_1 is felt at and amplified by Q_2 causing additional instability. For this reason, other types of coupling methods are used. One of these is RC coupling.

RC Coupling

RC, or resistance-capacitance, coupling prevents the problems associated with direct coupling, but is almost always used for medium to high frequency operation because the coupling capacitor, shown in Fig. 7-16, would block any dc or low frequency signal from one stage to the next. Notice in this schematic that voltage divider biasing is being used for these common-emitter circuits that are coupled together.

The charge on the coupling capacitor is equal to the potential difference between Q_1's collector and Q_2's base. Since the collector of Q_1 is more positive than the base of Q_2, any ac input signal at Q_1 will swing positive and negative at the collector of Q_1 but with a positive reference voltage. This means that the output signal of Q_1 is a replica of the input signal but riding on a positive

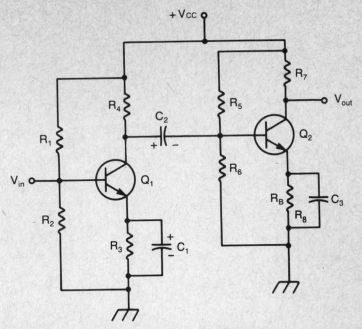

Fig. 7-16. RC coupled common-emitter amplifiers.

reference with respect to ground. The coupling capacitor charges up from ground through R_6, C_2, and through R_4 to V_{CC}. When C_2 discharges due to a changing I_C of Q_1, it again discharges through R_6 developing a varying voltage across the emitter-base junction of Q_2. This varying voltage at the base of Q_2 causes the base current of Q_2 to vary and in turn, I_C of Q_2 to vary as the input of Q_1 does.

Typically the coupling capacitor offers a very low impedance to the ac signal being transferred from one stage to the next. In most cases, this capacitor is electrolytic and of a value of 10 μF or greater.

The disadvantage of this circuit arrangement is that as the input signal frequency increases so does the reactance of the coupling capacitor, so there are upper and lower frequency limits with this type of coupling.

Impedance Coupling

Figure 7-17 is an example of impedance coupling. The only difference really between this circuit and that of Fig. 7-16 is that a coil, usually resistance wiring wrapped around a ferrite core, is

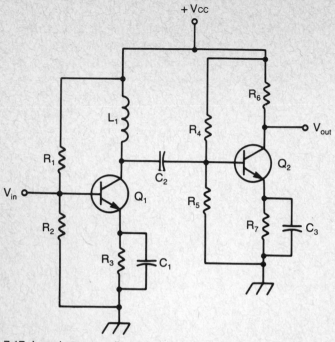

Fig. 7-17. Impedance coupling using two common-emitter amplifiers.

put in place of the collector resistor for Q_1. This inductor has a very low impedance to dc, but offers a high resistance to ac, so, in effect, the ac signal being coupled over to Q_2 sees a high value resistance in the collector circuit of Q_1. Since it offers a small resistance to dc, it drops a very small voltage across its windings adding to the efficiency of the operation of Q_1 by consuming very little power.

The biggest disadvantage of this type of coupling is that as the input signal frequency increases so does the relative impedance of the inductor. Since its impedance increases, so does the output voltage of Q_1 and therefore the overall voltage output gain of both stages. Of course, the total output will be limited by the beta of the transistor, but this type of coupling arrangement is usually limited to a single frequency to be amplified or a narrow band of frequencies that must be amplified by one stage and then passed on to another stage to be amplified.

Transformer Coupling

Although transformer coupling has many disadvantages, one

158

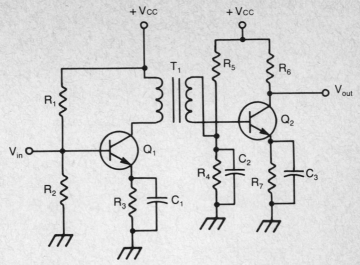

Fig. 7-18. Common-emitter amplifiers are cascaded in this illustration using transformer coupling.

being that it can only couple an ac signal, it is still used in some circuit applications. Figure 7-18 is a schematic of two amplifiers that have been coupled through a transformer. Any input changes are also felt across the primary winding of T_1 and inductively coupled to the secondary winding. Since the secondary winding is connected in series with the base of Q_2, these variations are felt there also, are amplified by Q_2, and taken from its collector.

The biggest advantage to this type of coupling is that the high impedance output of Q_1 can effectively be coupled to the low input impedance of Q_2 and in a sense the transformer serves as a buffer between the two. The low input impedance of Q_2 can look like a much higher input impedance to the output of Q_1 once a proper turns ratio has been achieved by adjusting the turns ratio of T_1's input and output windings.

CLASSES OF AMPLIFIER OPERATION

There are several different transistor classes of operation, which determine when current flows in the collector circuit of a transistor according to certain levels of base biasing. Classes that are most often used are A, B, AB, and C.

Class A

An amplifier biased to allow output current to increase and de-

159

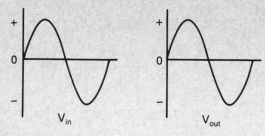

Fig. 7-19. In class A operation, the output signal waveform is a linear representation of the input signal waveform with no output distortion.

crease proportionately with increases and decreases in input signal current, and in a linear manner, is called class A operation. The easiest way to represent this linear operation is by viewing the illustration of Fig. 7-19. This output sinusoidal waveform, an amplification of a similar input signal, alternates positively and negatively, but never drops to zero during any alternation. The biasing keeps the output of the amplifier in the linear region of its operating range. If the transistor is biased so that its output current goes absolutely no higher, the transistor is saturated. Reducing the bias until the output current drops to zero or nearly so is called cutting off the transistor, or transistor cutoff. The transistor in class A operation operates so that it is always in the transition range between saturation and cutoff. In class A operation, minimum distortion is achieved.

Class B

When the transistor is biased so that output current flows for only one half of the ac input cycle, the transistor is operating as class B. Again, biasing allows an output current only when the input ac cycle has a positive alternation. This is shown in Fig. 7-20.

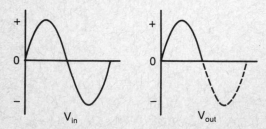

Fig. 7-20. In class B operation, only a portion of the input signal voltage is represented in the output.

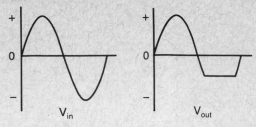

Fig. 7-21. The input signal is represented in the output when an amplifier is biased class AB.

Class AB

Operating the transistor so that its output current flows for less than one full ac cycle but more than one half of that cycle is called operating the transistor in class AB, shown in Fig. 7-21. This allows operation in class A with small signals, and operation in class B with larger signals.

Class C

Finally, class C operation is obtained by biasing the transistor so that its output current flows for less than one half of the ac input signal. This is achieved by biasing the transistor beyond cutoff so that the output current flows only during a portion of the positive alternation of the ac input signal.

Each of these classes of operation has a specific use in electronics. In most instances class A amplifiers are used primarily to amplify audio signals because of their minimum distortion and their good voltage gain characteristics. Class AB, B, and C are usually used in electronic circuits as power amplifiers. Some of these circuits are described in the next chapter.

Chapter 8

Typical Bipolar Amplifiers

The bipolar amplifiers presented thus far are general purpose amplifiers and cannot be used for certain applications without some modifications. There are a great many different kinds of amplifiers simply because there are a great many applications for these amplifiers. Some require linear reproduction of the input signal with emphasis on power rather than voltage amplification; more often, amplifiers are designed for a specific range of frequencies with minimum distortion. This chapter begins with the lower frequency amplifiers and progresses up to the higher frequency amplifiers.

DIRECT CURRENT AMPLIFIERS

Direct current amplifiers are not limited to amplifying only dc voltages and currents. They are designed to operate and pass frequencies from as low as dc to perhaps several thousand hertz.

In most cases very small dc voltages from such devices as pressure, light, and heat sensors must be brought up to a much higher level. The dc amplifying circuit brings the very small dc sensing output of these devices to a much higher level that is usable in an electronic circuit.

Since dc amplifiers are required to pass dc signals and very low frequency ac signals, they are designed to operate without transformers and capacitors in the signal path because these devices allow ac to pass but block dc. Coupling more than one stage of amplification must be done directly rather than through capacitive or inductive coupling.

Dc Amplifier Configurations

Figure 8-1 is an example of a basic dc amplifier. Notice that it uses voltage divider biasing and emitter feedback to prevent thermal runaway. This circuit responds well to dc signals and provides good voltage gain and is used for just this purpose in most cases. Remember, as long as there is no bypass capacitor around R_E, the voltage gain of the transistor is essentially equal to R_3 divided by R_E or:

$$A_v = \frac{R_3}{R_E}$$

This circuit is also useful for amplification of ac signals as well. The upper frequency limit of the circuit is a function of the beta of the transistor which begins to drop off after a certain frequency is reached. Low frequency transistors operate at frequencies of from dc to several thousand hertz and then show a sharp reduction in gain once a higher frequency signal is used as an input signal. Manufacturers usually specify the best operating frequency range

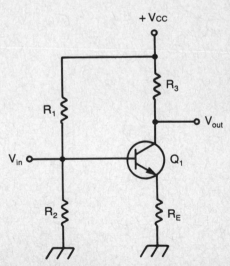

Fig. 8-1. A typical dc amplifier using voltage divider biasing is shown here.

for a particular transistor, especially for low power devices, and some manufacturers specify whether a device should be used for audio frequencies, intermediate frequencies (i-f), or even higher microwave frequencies. Figure 8-2 is an example of how a low power transistor is packaged and how the manufacturer specifies the recommended use for the device on the package.

Since the input signal source is connected effectively across R_2, (Fig. 8-1), the internal resistance of the signal source becomes part of the voltage divider biasing scheme affecting base current, and therefore, collector current. Therefore, this internal resistance must be taken into consideration when designing the input biasing for this circuit. In the same way, the load resistance, R_L, in the output circuit actually affects or alters the value of R_E; it too must be taken into consideration when biasing this dc transistor amplifier. This is important because the amplification of the device is affected when external components are connected to the input and output circuits of this amplifier.

Multistage Dc Amplifiers

As you may have guessed, one dc amplifier may simply not

Fig. 8-2. One of the many ways in which a manufacturer may package and specify the use of a particular transistor (John Sedor Photography).

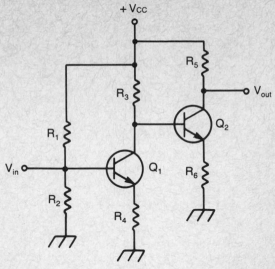

Fig. 8-3. Connecting transistors without the use of a passive component is called direct coupling.

provide enough amplification for a specific application. Figure 8-3 is an example of cascading dc amplifiers. Neither capacitive nor inductive coupling is used because capacitors block dc and inductors only pass an ac signal. Therefore in Fig. 8-3, Q_1 is directly coupled to Q_2. Although Q_1 is biased using voltage divider biasing, Q_2 need not be since the output of Q_1 is connected directly into the input of Q_2.

Voltage gain of this cascaded arrangement of Q_1 and Q_2 is found not by adding the gain of the first stage to the gain of the second stage but by multiplying their gains. If the first stage has a gain of 15 and the second stage also has a gain of 15, the overall gain provided by this multistage dc amplifier configuration is 225. Remember also that the common-emitter circuit of Q_1 phase shifts the input signal by 180° and so does the common-emitter circuit of Q_2. Phase shifting once, and then again twice, causes a total phase shift of 360°. Thus, the final output signal is an amplified reproduction of the initial input signal with no phase inversion.

In practice, a coupling resistor is connected between stages as shown in Fig. 8-4. The purpose of R_c is to prevent interaction of one stage from the other. Without R_c, the collector voltage of Q_1 would have to be lowered to a value that would correspond to the much lower value of voltage needed at the base of Q_2. This seriously reduces the efficiency and therefore the gain of Q_1. The

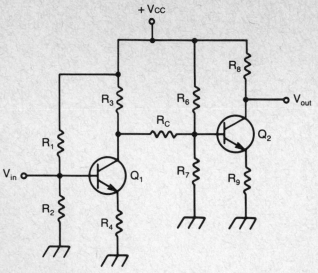

Fig. 8-4. A direct coupled multistage dc amplifier using a coupling resistor, RC.

overall gain of the multistage dc amplifiers would also be reduced. R_c then, is usually a high resistor value, preventing interaction between stages, but this also causes a loss of signal amplitude from the output of Q_1 to the input of Q_2. The overall voltage gain of this multistage dc amplifier is much lower than the product of each individual stage.

Another method of coupling which prevents interaction between stages and any significant reduction in the signal between Q_1 and Q_2 is shown in Fig. 8-5. The zener diode is reverse biased simply because the collector voltage potential of Q_1 is usually several volts higher than the base voltage of Q_2. A zener with the same voltage rating as the difference between the collector voltage of Q_1 and the base voltage of Q_2 must therefore be chosen for proper operation.

Notice that the zener diode and R_5 are really in series with each other and are therefore in parallel with the collector of Q_1 to ground. Since the voltage drop across the zener remains essentially constant, any voltage variations at the collector of Q_1 to ground are felt across R_5, which varies its voltage drop and therefore biasing on the base of Q_2 as well. The zener therefore provides isolation between stages with no significant signal loss between Q_1 and Q_2. Q_2 then amplifies the signal from Q_1 and once again the

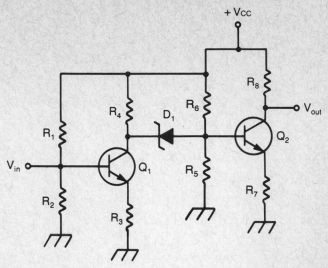

Fig. 8-5. Coupling is accomplished in this multistage dc amplifier using a zener diode.

overall voltage gain is essentially equal to the product of the gain of both stages.

Darlington Amplifiers

Another type of multistage transistor circuit configuration called a Darlington configuration is shown in Fig. 8-6. Either NPN or PNP transistors can be used in this arrangement which consists of two transistors connected together to function as a single unit. In fact, in most transistor packages you'll find both devices inside already connected together. The complete package has only three extended leads marked emitter, base, and collector.

The emitter current of Q_1 is also the base current of Q_2. Any variations of the input to Q_1 are felt at the base of Q_2. This causes the collector current of Q_2 to vary with input variations to Q_1. Q_1, therefore, directly controls the conduction of Q_2. The gain of the first stage is multiplied by the gain, or beta, of the second stage to determine the overall gain of the Darlington package. Darlington packages have extremely high voltage gain, sometimes as high as 1000 or 2000, and are used in applications where high gain is needed. However, too much of an input signal can quickly saturate Q_2 in the package. The collector of Q_1 is connected directly to Q_2, and the collector of Q_1 must be kept low enough to prevent too

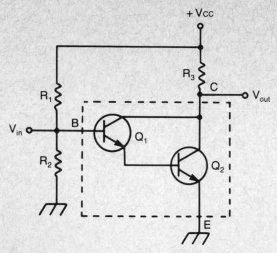

Fig. 8-6. Cascading amplifiers like this yields a multiplication of transistor gains rather than an addition and is called a Darlington configuration.

much base current from flowing in Q_2, causing that device to be saturated. The input signal is usually quite small.

AUDIO AMPLIFIERS

An audio amplifier is one of the most popular types of amplifiers in use today. Just think of how communications has progressed over the last several years, from stereo radio to stereo TV, and, in mobile communications, the car telephone. Audio frequency amplifiers, with a range of between 20 hertz and 20 k hertz, are sometimes also referred to as AF amplifiers.

Audio amplifiers are usually designed to operate class A, although class B amplifiers are also designed for some applications, and are designed for both voltage and power amplification.

POWER AMPLIFIERS

There are a number of different types of audio amplifiers used for power amplification. Power amplifiers usually act as the last stage of amplification in most amplifier configurations because power amplifiers usually require a signal that has already been amplified up to a specific level. The power amplifier amplifies the signal one final time before being delivered to a loudspeaker or other type of load.

Power amplifiers are rated in terms of rms watts, the same unit used in calculating power when current and voltage or resistance values are known. As an example, if the output of a power amplifier is 10 volts rms, and a 4 ohm load is connected to the output of the power amplifier, the amount of output power delivered to the load will be:

$$P = \frac{E}{R}$$

$$P = \frac{10}{4} = 25 \text{ watts, rms}$$

Rms watts are used because the output voltage across the load is measured as an rms value.

Figure 8-7 is an example of a single stage, or single-ended, power amplifier and is biased for class A operation. The damping factor of the amplifier is of concern here in power amplifier design. Basically, the output impedance of the amplifier should be as low as possible so that any load variations will not affect the output voltages of the amplifier. The damping factor is a ratio of these

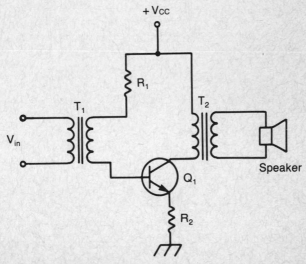

Fig. 8-7. A typical audio power amplifier with input and output transformers.

impedances and is found with the following equation:

$$D = \frac{Z_L}{Z_{out}}$$

D = Damping factor
Z_L = Load impedance
Z_{out} = Characteristic output impedance of the amplifier

The lower the load impedance, the lower the damping factor. Load impedance can be caused by wire length and size from amplifier to load (such as a speaker) and the resistance of the terminal connections. The lower the value of these items, the lower the damping factor and the better the sound from the audio power amplifier.

The circuit of Fig. 8-7 provides excellent gain but also has a relatively high damping factor. Also, the dc current flowing through the primary of T_2 along with the ac signal current could saturate the magnetic core of the output transformer, if the transformer is not large enough. In most instances this design, requiring bulky and expensive transformers, is exchanged for a more popular circuit configuration shown in Fig. 8-8.

Push-Pull Amplifiers

This arrangement is called push-pull because one amplifier is

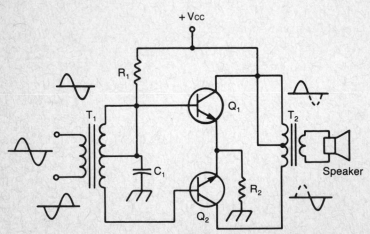

Fig. 8-8. A typical push-pull power amplifier using input phase splitting.

170

conducting while the other is cut off, depending on the alternation of the input signal waveform. In effect, the input signals must be complementary to each other (one signal is inverted as compared to the other signal). The input is a transformer or transistor circuit that splits the input signal and phase inverts one of the resulting waveforms so that the inputs to the bases of Q_1 and Q_2 are out of phase. The center tap of the input transformer, T_1, is connected to ac ground through capacitor C_1.

Push-pull amplifiers are usually biased class AB or B. This means that with no input signal, each transistor is still conducting slightly. Q_1 conducts during the positive alternation of the input signal while Q_2 remains cut off. On the negative alternation of the input signal, Q_1 is cut off while Q_2 conducts. The combined waveshape at the output of T_2 has nearly the same waveshape as the original input signal. Resistor R_2 allows current to flow through the emitter and up through the collector of Q_1 when the base of Q_1 is more positive than the base of Q_2. The current flow continues down through the upper half of T_2 and back to Vcc causing what is called a push in T_2. Upon conduction of Q_2 base current, from ground through R_2, through Q_2, out of Q_2's collector, and up through the lower half of T_2, the output now exhibits the pull of the push-pull operation of this circuit because this source of pulling is caused by current reversal in the output of T_2. In the process, although class AB or B is used, little distortion is caused in the output because the top and bottom halves of this type of amplifier are mirror images of each other.

Complementary Amplifiers

Another type of push-pull amplifier that eliminates the need for input and output transformers is called the complementary push-pull amplifier, shown in Fig. 8-9. Here, one of the transistors, Q_2, has been replaced with a PNP transistor. As the input signal becomes more positive, Q_1 conducts and Q_2 is cut off. When the input signal becomes negative, Q_1 is cut off and Q_2 conducts. The output from these emitter follower or common-collector amplifiers is coupled to the output load through a capacitor. The capacitor is used to block dc present at the emitters of Q_1 and Q_2.

D_1 and D_2 are used for several reasons. The first is to make sure that the emitter-base junctions of Q_1 and Q_2 together remain constant at 1.4 volts (0.7 volts for each junction). They are also used to prevent thermal runaway, where the collector current of the

171

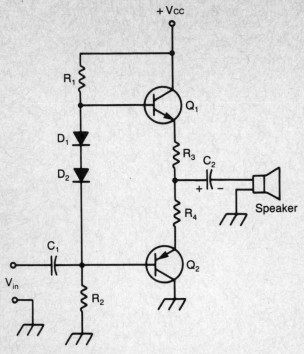

Fig. 8-9. A complementary push-pull power amplifier.

transistor rises to a critical point due to a temperature increase of the device as it conducts more and more collector current. D_1 and D_2 are placed physically close to Q_1 and Q_2. As the temperature in Q_1 or Q_2 increases, so does the current through them. When the diode temperature increases, so does the current through them. The voltage across them decreases, reducing the voltage difference between the bases of Q_1 and Q_2, lowering base current and returning the current through Q_1 and Q_2 back to normal. Of course, R_3 and R_4 serve the same function of thermal stability.

Quasi-Complementary Amplifiers

In any push-pull operation, the characteristics of the transistors must be evenly matched, especially as the power of the devices increases. A quasi-complementary amplifier operates like a complementary amplifier configuration, but does not require high power complementary output transistors as the output circuitry to provide the needed power and gain. Instead, the Darlington configuration is used to provide the needed gain, shown in Fig. 8-10.

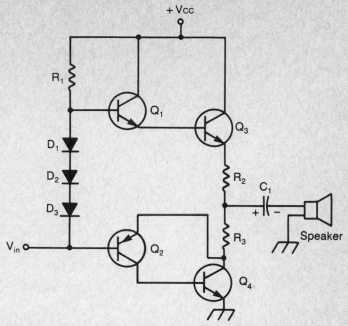

Fig. 8-10. A quasi-complementary audio power amplifier using two Darlington configurations.

Q_1 and Q_3 are connected in the Darlington fashion. Q_2 and Q_4, although they are PNP and NPN transistors respectively, still function like a Darlington configuration. The purpose of two NPN power transistors in the outer stage with the bottom arrangement containing only one PNP transistor is that in most cases NPN power transistors as power output devices are cheaper than PNP power transistors. The first two transistors, Q_1 and Q_2, are used as drivers for Q_3 and Q_4. The more positive the input signal, the more active are transistors Q_1 and Q_3 and the more negative the input signals, the more active are transistors Q_2 and Q_4. And, of course, there are three diodes instead of two, used for stabilization, because there are three transistor emitter-base junctions, Q_1, Q_2, and Q_3.

VOLTAGE AMPLIFIERS

Voltage amplifiers are usually the first stages of amplification of an input audio signal source before being sent to a power amplifier for final delivery to the load, such as a loudspeaker. The power amplifier supplies the high output signal current for driving the loudspeaker.

173

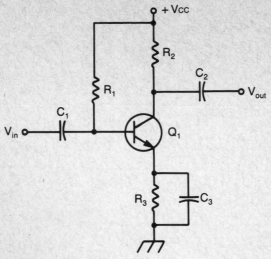

Fig. 8-11. A typical basic audio voltage amplifier using an NPN transistor.

A basic audio voltage amplifier is shown in Fig. 8-11. This is a common-emitter circuit using emitter feedback R_3, while C_3 serves to bypass the ac input signal which could cause degeneration of the transistor gain.

The input signal is applied to input capacitor C_1, but in reality is applied between the transistor's base and circuit ground. The output is taken from the collector of Q_1, more precisely between the collector of Q_1 and circuit ground, and passes through the coupling capacitor C_2 to the output. The frequency response of this amplifier is somewhat interesting and drops off at both the high and low ends as well, shown in Fig. 8-12. The decrease in

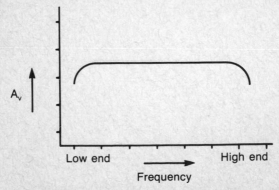

Fig. 8-12. Frequency response of the audio voltage amplifier of Fig. 8-11.

amplification at the high end is due to the beta of the transistor which decreases as frequency increases. This varies depending on the type of transistor used. The low end frequency causes a loss in gain also, but this is due to the capacitive reactance, X_C, of each capacitor at the low frequencies of the input signal voltage.

VIDEO AMPLIFIERS

There are frequencies, other than audio frequencies, that must also be amplified. Since audio frequencies are usually limited to about 20 kHz, other high frequency amplifiers must be designed that can amplify frequencies up to approximately 6 MHz, suitable in radar or in television for amplifying picture information. These are also wideband in nature, from about 10 Hz at the lower end to about 6 MHz at the upper end. Another way of saying this is that video amplifiers must respond, or have a relatively flat frequency response, from 10 Hz to 6 MHz.

Frequency Response

Figure 8-13 shows the typical frequency response of an amplifier with an audio and video bandwidth. An RC coupled amplifier is suitable for this type of frequency response because its frequency response is flat in the middle area of the frequency range. However, the lower and upper ends of the frequency response curve deserve a little more attention because it's rather difficult to design an amplifier with as sharp a frequency response curve.

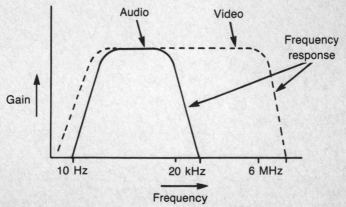

Fig. 8-13. Typical frequency response of an amplifier with audio and video bandwidths.

175

A number of different factors affect frequency response of a video amplifier. Remember, the junction of a transistor behaves very much like a capacitor. The depletion region of the transistor junction acts like a dielectric while the semiconductor materials on either side act like the plates of a capacitor. As the junction is forward biased and the depletion region decreases, the junction presents greater capacitance. Even without forward bias, junction capacitance is still a problem at very high frequencies. These capacitances are shown in Fig. 8-14.

As the emitter-base junction is forward biased it presents a higher capacitance. And as the collector-base junction is reverse biased it also introduces capacitance, but at a somewhat lower level than that of the forward biased emitter-base junction. Most manufacturers try to keep these shunt capacitances to a very low level. All of these internal capacitances have an effect on the high frequency gain of the transistor.

Another type of capacitance that exists between the transistor's collector-base junction causes a phenomenon known as the Miller effect. This capacitance causes degenerative, or negative, feedback between collector and base junctions and is in series with the internal resistance between the transistor's collector-base junction, shown in Fig. 8-15.

At the higher frequencies, X_C of C_{CB} decreases, reducing the feedback path reactance between collector and base, therefore increasing negative feedback, resulting in a loss of gain. Miller ca-

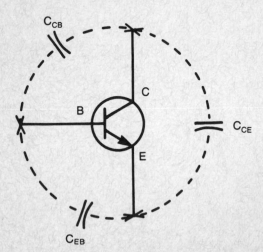

Fig. 8-14. Junction capacitances found in a typical bipolar junction transistor.

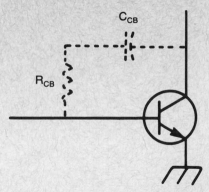

Fig. 8-15. A type of junction capacitance and resistance between collector and base produces a phenomenon called the Miller Effect.

pacitance can be found with the following equation:

$$C_M = C_{CB} (\beta + 1)$$

C_M = Miller effect
C_{CB} = Collector-base junction capacitance
β = Circuit gain

As you can see, as beta increases so does the Miller effect. Miller effect therefore affects the frequency response of the transistor at higher frequencies because this negative feedback from collector to base increases due to a reduction in capacitive reactance in the feedback path from collector to base.

Shunt capacitances are drawn in as shown in the RC coupled network of Fig. 8-16. Here, the effective output capacitance of Q_1 is shown along with the input capacitance of Q_2. Mathematically,

$$C_{out} = C_{CE} + C_S$$
and,
$$C_{in} = C_M + C_{BE} + C_S$$
$$\text{or } C_{in} = C_{CB(\beta + 1)} + C_{BE} + C_S$$
and
$$C_T = C_{in} + C_{out}$$

C_T = Total shunt capacitance between the output of Q_1 and the input of Q_2
C_{CE} = The internal capacitance between collector and emitter
C_{CB} = The internal capacitance between collector and base
C_S = Stray capacitance of wiring in the circuit used to connect components

177

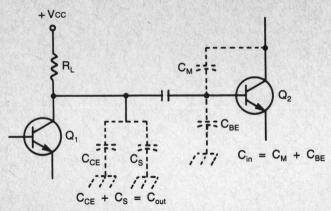

Fig. 8-16. Shunt capacitance can be viewed as imaginary capacitors as shown here, between RC coupled stages.

The total shunt capacitance, that capacitance between the output of Q_1 and the input of Q_2 in Fig. 8-16, can be as high as 200 pF. What this means simply is that some of the higher frequency ac signal voltages will be shunted to ground. In fact, a cutoff frequency called the high frequency cutoff, F_{CO}, is reached when the output voltage of Q_1 has decreased to 70.7% of its maximum value. This occurs in effect, when $X_{CT} = R_L$. F_{CO} can be calculated from the following equation given 10 k ohms for R_L and 200 pF for C_T in Fig. 8-17:

$$F_{CO} = \frac{1}{2 \pi R_L C_T}$$

$$F_{CO} = \frac{1}{2(3.14)(10 \times 10^3)(200 \times 10^{-12})}$$

$$F_{CO} = \frac{1}{6.28(2000 \times 10^{-9})}$$

$$F_{CO} = \frac{1}{12560 \times 10^{-9}} = \frac{1}{1.256 \times 10^{-7}}$$

$$F_{CO} = 79.6 \text{ kHz}$$

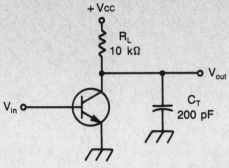

Fig. 8-17. In this circuit, shunt capacitance, C_T, has a direct effect on the upper frequency response of the amplifier.

To increase the cutoff frequency, and in effect the frequency response of the circuit to a much higher frequency, R_L can be reduced from 10 k ohms to 1 k ohm. This increases F_{co} substantially up to about 800 kHz, but with a significant loss in voltage gain. Gain is sacrificed for greater frequency response until a point is reached where gain is practically non-existent and the 6MHz bandwidth still has not been achieved.

A better method for extending the frequency response of the video amplifier is with the use of peaking coils, shown in Fig. 8-18. There are three different ways in which a peaking coil may be connected: in a shunt mode, in a series mode, and in a series-shunt mode of operation. In Fig. 8-18 the shunt mode is used. In effect, the peaking coil is in parallel with C_T and nullifies the shunt capacitance. L_{SH} resonates with C_T, increasing the output impedance

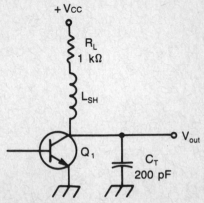

Fig. 8-18. In this configuration, shunt peaking is used to increase frequency response.

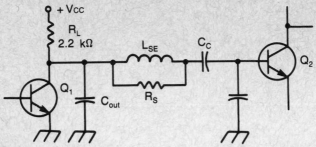

Fig. 8-19. Increasing frequency response of a multistage amplifier using series peaking.

of Q_1 and therefore the gain at the higher end of the frequency response curve. In fact, this method can increase the upper end of the frequency response curve by a factor of 2.

Series peaking is shown in Fig. 8-19. Inductor L_{SE} effectively isolates the input and output capacitances of Q_1 and Q_2. This allows an increase in resistor R_L, so the gain can once again go back up. Resistor R_S is referred to as a swamping resistor and is actually the core around which the inductor is wound and is used to prevent ringing of the coil. In other words, it has a damping effect on L_{SE}.

Figure 8-20 is an example of the combinational use of a series peaking coil and a shunt peaking coil. In this circuit configuration L_{SE} opposes the input capacitance of Q_2 while L_{SH} is used to nullify the output capacitance of Q_1. The bandwidth, with both of these devices used in this manner now extends the bandwidth of the video amplifier to the required 6 MHz for video amplification.

Finally, a typical video amplifier is shown in Fig. 8-21. Notice the combination series and shunt peaking coils to extend the frequency response, R_4 degenerative feedback used for adjustment of gain, R_5 for thermal stability, C_3 for bypassing the ac components to ground, and voltage divider biasing provided by R_1 and R_3.

RF AND I-F AMPLIFIERS

Rf or radio frequency and i-f or intermediate frequency amplifiers are a necessary part of wireless communications from one point to another whether fixed or mobile.

Rf amplifiers are basically amplifiers that are used in the first stages of amplification in radio and television receivers. This is because they are high frequency amplifiers that must take these

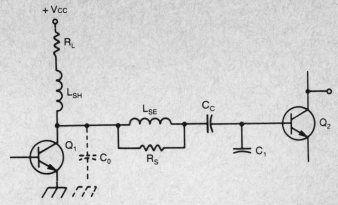

Fig. 8-20. The combinational use of both a shunt and series peaking coil for added frequency response.

high frequency transmitted signals, amplify them, and then send them on for further processing by the radio or television receiver. Amplification is necessary because these transmitted signals, by the time they reach the receivers, are very small indeed. Rf

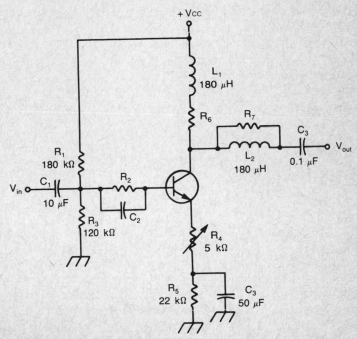

Fig. 8-21. A typical stage of a video amplifier using peaking coils, thermal stability, and voltage divider biasing.

amplifiers therefore are used to amplify very small high frequency signals of only a few hundred microvolts to more useful high frequency signals for further processing.

These signals are then sent to i-f amplifiers which take the variable rf frequencies and change them to a fixed frequency signal. The major difference between the two is that the rf amplifier is tunable over a very wide frequency range while the i-f amplifier is set at a fixed intermediate frequency, usually 455 kHz in the case of AM radio. The main aspect that these two types of amplifiers have in common is that they use a tuned amplifier for amplification and selectivity. These points will be covered in the topics to follow.

Tuned Amplifiers

Tuned amplifiers allow only a specific band of frequencies to pass while rejecting all other frequencies above and below that specific band. It attains this objective by using adjustable resonant circuits at the input and output of the circuit, shown in Fig. 8-22. As you can see, the inductors used in the resonant circuits are actually part of the input and output coupling transformers.

At the resonant frequency in the input circuit, maximum voltage is developed across the base of Q_1. This particular frequency (actually a very narrow band of frequencies) is amplified and developed across the tuned circuit consisting of L_2 and C_2. The natural frequency at which these tuned circuits resonate and develop the proper input and output frequencies is determined by the following equation:

$$F_O = \frac{1}{2 \pi \sqrt{LC}}$$

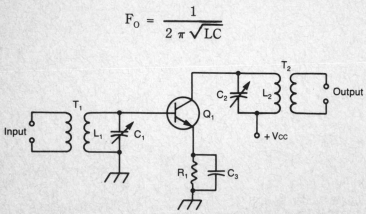

Fig. 8-22. A tuned amplifier with adjustable input and output resonant circuits.

These resonant circuits produce a response curve like the one shown in Fig. 8-23. It's important to remember that impedance of a parallel resonant circuit is maximum at resonance. This chart therefore shows that as impedance is maximized, so is output frequency. Below the half-power points circuit response is simply too low to produce a useable output.

Selectivity is an important factor here too. The ability of the circuit to select one frequency and reject all others is called selectivity. The narrower the bandwidth, the greater the selectivity. Bandwidth is also a function of the Q of the circuit (the quality factor). The higher the circuit Q, the narrower the bandwidth and the greater the selectivity. Also, the higher the Q, the higher the impedance of the resonant circuit. All of these factors are related and can be expressed as follows:

$$BW = \frac{F_O}{Q}$$

In most rf amplifiers however, a wide bandwidth is required but so is a high gain. It might seem that the two are incompatible, but a compromise can be reached by introducing particular coupling techniques. In effect, transformer coupling is used and optimum coupling is achieved when there is a relatively flat response curve from the center frequency and out to the half-power

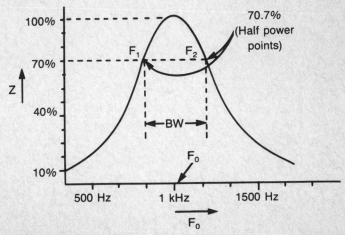

Fig. 8-23. Frequency response curve showing how BW is measured using the half-power points.

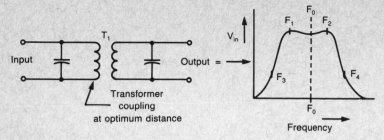

Fig. 8-24. Optimum transformer coupling produces the ideal output waveform.

points, and then a sharp drop (steep skirts) on either side of F1 and F2, shown in Fig. 8-24.

Figure 8-25 shows an rf amplifier and input and output tuned circuits while Fig. 8-26 shows the input and output waveforms of these circuits.

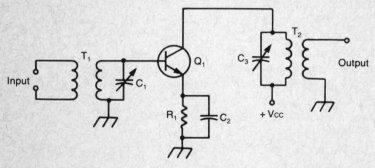

Fig. 8-25. A typical rf amplifier using tuned input and output tank circuits.

Rf Amplifiers

A typical rf amplifier is shown in Fig. 8-27. Notice C_1 and C_3 are gang tuned. This means that they are tuned to the same frequency at the same time. This is the type of amplifier that might be found in the front end of an AM radio receiver.

AGC bias at the input to Q_1 maintains a constant gain for variations in the strength of the input signal by conversely varying the bias on the base of Q_1 as the input signal strength varies. The load circuit of C_3 and T_1 provides high voltage gain at the desired resonant circuit frequency. The purpose for tapping into T_1 is so that a good impedance match may be obtained between the transistor collector and the output transformer. This circuit operates as class A.

Another class A operated rf amplifier is shown in Fig. 8-28.

184

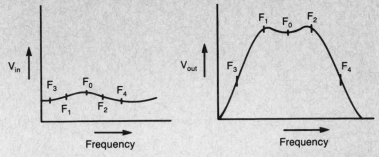

Fig. 8-26. Input (left) and output (right) waveforms of the rf amplifier shown in Fig. 8-25.

In this amplifier, the arrows are on the ends of inductors L_1, L_2, and L_3. These inductors, used for selectivity of the signal to be amplified and passed through, are switchable, not with each other, but with other inductors elsewhere in the television vhf tuner where this circuit is found. They are switched out, all at the same time, and the other inductors put in their place, when the channel selector knob is rotated for the TV to receive a new channel, or range of frequencies.

Once again, bias is controlled by a voltage feedback signal called automatic gain control, AGC, that controls the voltage level on the base of Q_1. A filter network made up of R_3 and C_6 keeps rf out of the power supply shown as + 10 volts. Notice also that the symbol for the transistor has dotted lines around it. This indicates rf

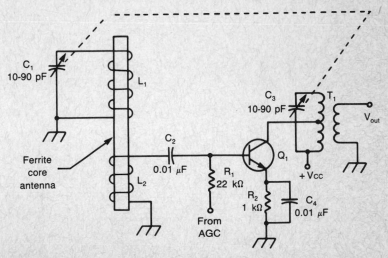

Fig. 8-27. A typical front end (1st stage) rf amplifier found in a radio receiver.

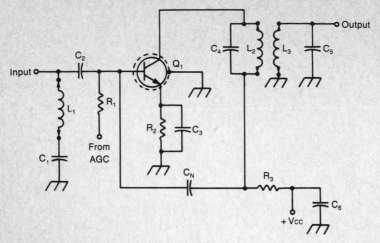

Fig. 8-28. A typical rf amplifier with selectable tuning coils.

shielding and keeps stray rf signals from interfering with the operation of the transistor.

I-f Amplifiers

In most receiver systems, the variable incoming rf signal is changed to a lower fixed frequency that has the advantage of being optimized more easily than a wide range of frequencies varying across a certain frequency range. I-f amplifiers can then be used to easily increase gain and control more easily the selectivity and sensitivity of the overall receiver. Most i-f amplifiers do not give sufficient gain in themselves and are therefore cascaded for better amplification performance.

I-f amplifiers, regardless of the i-f signal being received, amplify only a specific frequency. In AM radio receivers, this is 455 kHz. In practice the AM radio i-f amplifier usually has a bandwidth of 10 kHz so that it passes frequencies from 450 kHz to 460 kHz, with 455 kHz being the center frequency. Of course, the skirts of the response curve are not so sharp that frequencies slightly above and below do not pass as well. This 455 kHz is standard only in AM radio. In FM radio the i-f frequency is usually 10.7 MHz with a 150 kHz bandwidth, while in television from channels 2 through 83, the i-f varies from 41 to 47 MHz with a bandwidth of 6 MHz. The i-f has a broad range in television because the television signal has both audio and video components; there are separate i-f amplifiers for each.

186

Chapter 9

Operational Amplifiers

Entire books have been written on the seemingly endless uses of one of the most popular amplifiers in use today, the operational amplifier. Due to its variety of uses, the operational amplifier, or op amp, has found its way into most circuits, systems, and electronic equipment in use today. But this was not always the case. Not until manufacturing techniques introduced the integrated circuit, did the operational amplifier become so popular. It became a natural for the integrated circuit because transistor matching is critical for the transistors which make up the differential front end of the device.

Although operational amplifiers in integrated circuit form are relatively inexpensive, they contain a number of components that in discrete form had previously prohibited their use because of the size of the circuitry, cost of components, and difficulty in matching the characteristics of the transistors used in the differential section of the circuitry. Inexpensive integrated circuit manufacturing techniques change all that. All of these components are now mounted on one chip and as easy to use as a single transistor itself.

Operational amplifiers are now used as comparators, oscillators and function generators, active filters, Norton amplifiers, timers, in regulated power supplies, and even in phase-locked loop circuitry. This chapter begins with the input circuitry found in the op amp and proceeds to show some of the applications of op amps.

DIFFERENTIAL AMPLIFIERS

The input stage to most operational amplifiers in use today is the differential amplifier. Prior to the development of the integrated circuit, discrete components had to be used which required characteristic matching. The chief disadvantage of using such transistors is not only their cost, since at least two must be used, but the difficulty of precisely matching the characteristics of discrete transistors. Integrated circuit techniques allow the precise matching of transistors because the doping and even the geometry of the devices can be very evenly matched. Temperature changes are also controlled and matched more precisely between transistors in an IC because of their proximity or closeness to one another.

A number of different types of differential amplifiers are discussed beginning with the simplest.

Single-Input, Single-Output Amplifier

Figure 9-1 is a differential amplifier that contains a single input and a single output. In this configuration a positive input signal on the base of Q_1 causes a similar signal on the emitter of Q_1 due to emitter follower action. This causes the emitter of Q_2 to be positive, resulting in a positive amplified signal on the collector, and output, of Q_2.

Single-Input, Differential Output Amplifier

Figure 9-2 takes advantage of the differential transistor amplifier of Fig. 9-1. Here, an input signal at the base of Q_1 is amplified and inverted at the collector of Q_1, or V_{out1}. The signal on the emitter of Q_1 is fed to the emitter of Q_2 where it is

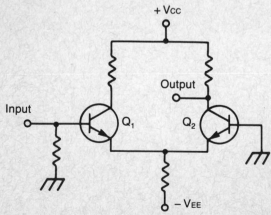

Fig. 9-1. A single-input, single-output amplifier.

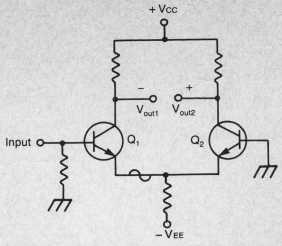

Fig. 9-2. A single-input, differential-output amplifier.

amplified and taken from the collector of Q_2, called V_{out2}. Still, the full advantage of the differential amplifier is not being used in this type of configuration.

Differential Input, Differential Output Amplifiers

In Fig. 9-3, full advantage of the characteristics of the differential amplifier are taken into consideration. In this circuit there are two inputs. The output is the amplified difference between the two input signals; thus the name differential or differencing amplifier.

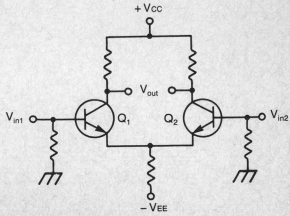

Fig. 9-3. A differential-input, differential-output amplifier.

189

In practice, the two input signals are 180° out of phase, making their difference output twice the amplitude of either input signal. Very simply, if the input at V_{in1} is 50 mV positive and the input at V_{in2} is 50 mV negative, the difference between the two is $50 - (-50) = 100$ mV. This type of arrangement provides a large output voltage developed across the collectors of Q_1 and Q_2 because the signals at these collectors are 180° out of phase from each other. The output is normally taken between the output of either collector and ground. A requirement of proper differential amplifier operation is a constant current source so that a constant current may be supplied from voltage source $-V_{EE}$ to each transistor. Practical current sources are considered next.

Constant Current Sources

An ideal constant current source is one from which a constant current may be supplied in spite of changes in load current. Figure 9-4 is an example of a basic constant current source. Since the emitter-base junction drops a constant voltage of 0.7 volts, the voltage drop across R_E remains constant because D_1 keeps the entire voltage drop from base to ground at 5.6 volts. R_E drops 4.9 volts and the emitter current is 4.9 mA. For all practical purposes, if points 1 and 2 are shorted, the collector current is also 4.9 mA. Even a 1 k ohm resistive load connected across points 1 and 2 does not significantly affect the current flow from ground, up through the transistor, and to V_{CC}. If the load is made large enough, current begins to be impeded.

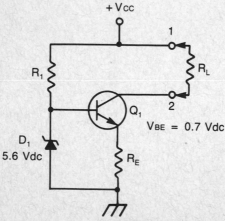

Fig. 9-4. A simplified constant current source.

A better constant current source is shown in Fig. 9-5. Here V_{BE} is equal to 0.7 volts. Since the base is grounded and the emitter must be negative in reference to the base for the emitter-base junction to be forward biased, the voltage on the emitter must be -0.7 volts. V_{RE} is therefore 10.7 volts $-$ 0.7 volts, or 10 volts. I_E therefore equals 10 V/10 k ohms which equals 1 mA. And since base current is negligible, collector current is also equal to 1 mA. The load resistor must not drop more than 15 volts. If it does, the voltage at the collector of the transistor is equal to zero volts, the same as the voltage at the base, and the collector-base junction will no longer be reverse biased. A load of up to 15 k ohms can be placed between points 1 and 2 without affecting the current flowing from the negative voltage source, up through the transistor, through the load, and to V_{CC}. As an example, if R_L equals 10 k ohms:

$$V collector = IR$$
$$V collector = 1 \times 10^{-3} \times 15 \times 10^3$$
$$V collector = 15 \text{ volts}$$

Figure 9-6 shows a differential amplifier with the symbol for the constant current source and Fig. 9-7 shows a practical common current source supplying current in equal amounts to each transistor of the differential amplifier in this arrangement. For the sake of clarity, electron flow rather than conventional current flow is being used.

DIFFERENTIAL AMPLIFIER CHARACTERISTICS

Several characteristics of the differential amplifier make it

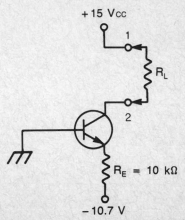

Fig. 9-5. A constant current source improved over Fig. 9-4.

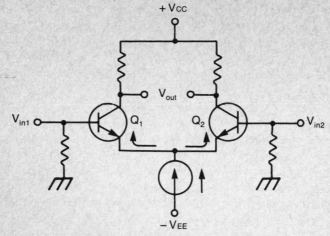

Fig. 9-6. A differential amplifier with a constant current source and its symbol.

suitable for a great many applications. One is its ability to amplify only those signals that present a difference in potential to the input of the differential amplifier. It can only do this if the common current source is consistent in its delivery of current. The reasoning for this and other characteristics is discussed next.

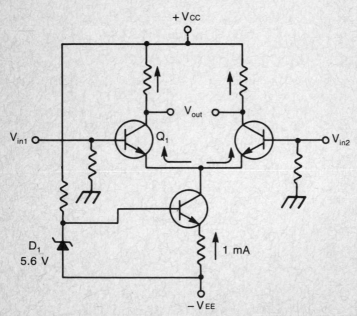

Fig. 9-7. A differential amplifier with a more practical constant current source.

Common-Mode Rejection

The differential amplifier has the ability to reject signals that appear exactly the same in phase and amplitude at the inputs to the device. These are called common mode signals. As an example, V_{out} is equal to the difference between the two input signal voltages. If V_{in1} equals V_{in2} then their difference is equal to zero, as shown in Fig. 9-8. As V_{in1} swings positive, Q_1 conducts more, but so does Q_2 as its input swings positive. Since the current source provides a constant current of 1 mA, both transistors conduct 0.5 mA, no more and no less, once they are turned on. Therefore the output voltage, V_{out}, does not change and remain at zero volts. The inputs are basically of the same phase and amplitude causing the same collector voltage on each transistor and therefore no potential difference at the output. This is the reason for the importance of evenly matched transistors and a current source that provides a constant source of current.

Differential Inputs

When the inputs are different or opposite in phase and amplitude, the differential amplifier amplifies these differences. Again, using Fig. 9-8, if the input to Q_1 goes positive while the

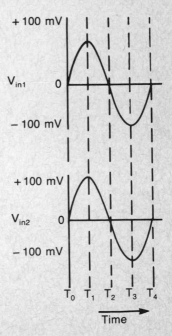

Fig. 9-8. Comparison of input signal voltages to a differential amplifier over a period of one complete input cycle.

193

input to Q_2 goes negative, Q_1 conducts more than Q_2, thus drawing more current than Q_2 from the constant current source. The current still remains a total of 1 mA, but Q_1 may conduct 0.7 mA while Q_2 conducts 0.3 mA. This decreases the collector voltage of Q_1 and increases the collector voltage of Q_2. There is now a difference in potential and therefore a voltage output from Q_1 and Q_2.

Common Mode Rejection Ratio

One of the best ways to measure just how well a differential amplifier rejects common mode signals and amplifies the differential signal is called the common mode rejection ratio (CMRR). This is the ability of the differential amplifier to amplify the difference signal while rejecting the common mode signal. To find the CMRR requires some simple calculations. Following them in a particular sequence gives you the CMRR.

The gain of the difference signal is first found from the following equation:

$$A_D = \frac{\Delta V_O}{\Delta V_D}$$

A_D = Difference gain
ΔV_O = Change in output voltage
ΔV_D = Change in differential input voltage

Next, the common mode gain is found:

$$A_{CM} = \frac{V_O}{V_{CM}}$$

A_{CM} = Common mode gain
V_O = Change in output voltage
V_{CM} = Change in common mode input voltage

Finally, the CMRR is found as follows:

$$CMRR = \frac{A_D}{A_{CM}}$$

As an example, find the CMRR if a change of 10 mV at the differential input causes a change of 1 volt at the output and a common mode input change of 100 mV causes an output change of 1 mV. To solve:

$$A_D = \frac{V_O}{V_D} = \frac{1\ V}{0.010\ V} = 100$$

$$A_{CM} = \frac{V_O}{V_{CM}} = \frac{0.001\ V}{0.100\ V} = 0.01$$

$$CMRR = \frac{A_D}{A_{CM}}$$

$$CMRR = \frac{100}{0.01} = 10,000$$

OPERATIONAL AMPLIFIERS

Differential amplifiers are used as the first stage in a very special type of amplifier called an operational amplifier, or op amp. Op amps are capable of providing very high gain, possess a very high input impedance, and a very low output impedance. Because of these characteristics, op amps are used in a great many applications as mentioned at the beginning of this chapter.

Since op amps are rarely found in discrete form any more, the IC op amp will be discussed only. Some of the advantages are less cost and smaller size over a conventional op amp constructed using discrete components. There are also op amps in IC packages that can be used with either a dual power supply or with a single supply. And most op amp IC packages contain at least two op amps, and some contain four. Figure 9-9 is a photograph of several kinds of op amps. As you can see, they come as 8 pin and 14 pin DIP (dual-in-line package). These particular op amps are the 741, one of the earlier and still popular dual supply devices; the 339 containing 4 op amps in one package; the 1458 op amp containing 2 op amps; and the TL082 op amp IC containing JFET op amps. TL082 op amp ICs, although having the same pin-out as the 1458, are not interchangeable with the 1458 because biasing

Fig. 9-9. Some typical operational amplifiers in IC form (John Sedor Photography).

arrangements are different with the JFET op amp than with the conventional bipolar transistor op amp.

Most op amps are composed of three stages, as shown in Fig. 9-10. The first stage is the input stage that has been already presented, the differential amplifier. The second stage is the voltage amplifier, usually composed of Darlington configurations, giving the op amp its extremely high gain. The last stage is a power amplifier output. To obtain a low output impedance, this final stage is usually an emitter follower operated class B. Since the input device is a differential amplifier, the op amp is capable of amplifying input signal frequencies down to dc as well as ac input signals.

Figure 9-11 is the symbol for the IC op amp. There are two inputs and one output. The inputs are labeled + and −, with the + input the noninverting input and the − input as the inverting input. With an input to the inverting input and the noninverting

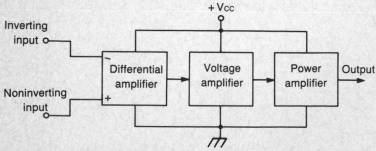

Fig. 9-10. The three basic stages of an op amp.

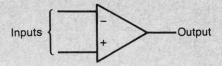

Fig. 9-11. The symbol for the op amp. Power supply connections are understood without drawing them in.

Inputs Output

Fig. 9-11. The symbol for the op amp. Power supply connections are understood without drawing them in.

input at ground, an amplified but inverted version of the input signal is seen at the output of the op amp. The same holds true for grounding the inverting input and applying an input signal to the noninverting or + input. The output is now an amplified version of the input signal without phase inversion. These are the two basic op amp circuit configurations and each of these will be covered next.

The Inverting Amplifier

Figure 9-12 is an example of the circuit arrangement used for the inverting op amp. Here, the input signal is fed to the inverting input through R_1. The noninverting input is tied to ground through R_2. R_f is a resistor which allows feedback to be taken from the output of the op amp and fed back to the input. This feedback signal will, of course, be 180° out of phase with the input signal and is negative, or degenerative, in nature. This nullifies the input signal and the input to the op amp is therefore zero volts, or virtual ground. If the input signal now increases to a positive 1 volt, the negative output of the op amp is fed back to the input to cancel the +1 volt input and keep the input to the op amp at virtual ground. Of course the feedback signal does not nullify the input signal completely— there is still about a microvolt or two at the input. If there were not, the op amp simply would not operate.

Since the input impedance of the op amp is extremely high, current from the input signal cannot flow into the noninverting input. In fact, current does not flow out of this input either. Current

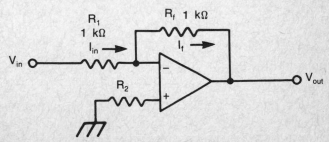

Fig. 9-12. An inverting op amp circuit arrangement.

flows through the input resistor, R_1, and around the op amp through the feedback resistor, R_f. Since the current flows through R_1 and the voltage at the input terminal is virtual ground, or zero volts, the input signal must be dropped entirely by R_1. The input current is simply:

$$I_{in} = \frac{V_{in}}{R_1}$$

I_{in} = Input current
V_{in} = Input signal voltage
R_1 = Input resistor

Since V_{out} is out of phase with V_{in}, a feedback current, I_f, is caused to flow through R_f from the input of the op amp to its output. I_f is therefore equal to:

$$I_f = \frac{-V_{out}}{R_1}$$

I_f = feedback current
$-V_{out}$ = output voltage 180° out of phase with V_{in}
R_f = feedback resistor

Since no current flows into or out of the input of the op amp, I_f must be the current that comes from I_{in}. Therefore, $I_{in} = I_f$. Substituting in this equation for I_{in} and I_f:

$$I_{in} = I_f$$

therefore:

$$\frac{V_{in}}{R_1} = \frac{-V_{out}}{R_f}$$

or:

$$\frac{V_{out}}{V_{in}} = \frac{R_f}{R_1}$$

Since gain, A, is equal to V_{out}/V_{in} then gain for this op amp must also equal:

$$-\frac{R_f}{R_1}$$

The minus sign indicates phase inversion. If R_f and R_1 are equal as shown in Fig. 9-12, then the gain of the op amp is equal to unity or 1. If V_{in} is equal to $+1$ volt dc, the current through R_1 is equal to:

$$I_{in} = \frac{V_{in}}{R_1} = \frac{1\ V}{1 \times 10^3\ ohms}$$

$$I_{in} = 1\ mA$$

I_{in} also flows through I_f, so $I_f = 1$ mA. V_{out} equals:

$$V_{out} = R_f(I_f)$$
$$V_{out} = 1 \times 10^3(1 \times 10^{-3})$$
$$V_{out} = 1\ volt$$

The circuit has unity gain. If the value of Rf is changed to 10 k ohms, the gain of the op amp circuit can still be found:

$$A = -\frac{R_f}{R_1} = \frac{10 \times 10^3}{1 \times 10^3} = -10$$

The gain of this circuit is -10. Going back to the original input voltage of $+1$ volt dc, the input current can be found as follows:

$$I_{in} = \frac{V_{in}}{R_1} = \frac{1\ V}{1 \times 10^3} = 1\ mA$$

This current must also flow through R_f to produce V_{out}. To find V_{out}:

$$V_{out} = I_f \times R_f$$

$$V_{out} = 1 \times 10^{-3} \times 10 \times 10^3$$
$$V_{out} = 10 \text{ volts}$$

Now, since this is an inverting amplifier, V_{out} must be negative, so V_{out} equals -10 volts. Since V_{in} equals $+1$ volt and V_{out} equals -10 volts, the circuit gain is equal to -10, as stated above. Since the input resistance to the op amp is extremely high and virtually all of the input voltage is dropped across R_1, R_1 determines the input resistance to this op amp configuration.

The Noninverting Amplifier

In the case of the noninverting amplifier shown in Fig. 9-13, the signal input is applied to the plus, or noninverting, input. Feedback is still applied to the noninverting input. This time however, the input resistance of the op amp is really the input impedance of the noninverting input of the op amp. Also, V_{out} is now in phase with V_{in}.

R_f and R_1 form a voltage divider or V_{out} being fed back to the input of the op amp. This feedback voltage, V_f, is felt between R_f and R_1. This feedback voltage is in phase with V_{in}. Even though V_f is in phase with V_{in} it still opposes the input signal V_{in}. In effect, when V_{in} swings positive, V_{out} swings significantly more positive, and thus so does V_f. This tends to offset the increase in V_{in} making V_f almost equal to V_{in}. For all practical purposes then, $V_{in} = V_f$.

To find the gain of this noninverting op amp circuit:

$$A = \frac{V_{out}}{V_{in}}$$

$$\text{or } A = \frac{V_{out}}{V_f} \text{ (since } V_{in} = V_f)$$

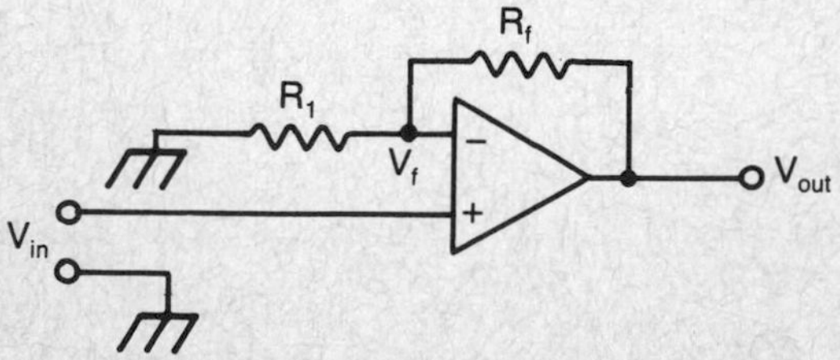

Fig. 9-13. An op amp in the noninverting configuration.

Since R_f and R_1 form a voltage divider and V_f is determined by V_{out} and R_f and R_1, V_f may also be determined using the voltage divider formula:

$$V_f = \frac{R_1}{R_f + R_1} (V_{out})$$

$$\frac{V_f}{V_{out}} = \frac{R_1}{R_f + R_1}$$

$$\text{or} \quad \frac{V_{out}}{V_f} = \frac{R_f + R_1}{R_1} = \frac{R_f}{R_1} + \frac{R_1}{R_1} = \frac{R_f}{R_1} + 1$$

$$\frac{V_{out}}{V_f} = \frac{R_f}{R_1} + 1$$

Earlier it was stated that:

$$A = \frac{V_{out}}{V_f}$$

Substituting then, gives:

$$A = \frac{R_f}{R_1} + 1$$

The gain of the noninverting op amp circuit is again a function of the ratio between feedback resistor R_f and input resistor R_1. Incidentally, both configurations are considered closed loop circuits because the feedback loop is closed, in this case, with a feedback resistor. Also, although V_f theoretically equals V_{in}, there is always a slight difference of potential between these two voltages. This difference causes the very small input current which is much smaller than the input current of the inverting op amp circuit. Input current of this circuit is not caused by V_{in} being dropped by R_1. This causes the input resistance of the noninverting amplifier to look much higher than the input impedance of the inverting amplifier.

Op Amp Bandwidth

As with any amplifier, gain and bandwidth are sometimes interrelated. Such is the case with the operational amplifier. In fact, in the open loop mode of operation where a feedback path is left open, the frequency response of the op amp is severely limited. In most op amps, there is an inverse linear proportion between gain and frequency response. As frequency increases, gain decreases, to a point where unity gain is achieved in the open loop mode when the frequency response of the op amp reaches about 1 MHz for the 741 op amp. The useful operating frequency of most op amps is actually closer to dc in the open loop mode of operation.

A way to increase the frequency response, and thus the bandwidth, of the op amp is by using degenerative feedback. Figure 9-14 shows a graph of the frequency response of an op amp in the open loop mode and that of an op amp using degenerative feedback to bring the gain down to 100 and the frequency response up from 10 Hz to 10 kHz. (This is a typical 741 curve.) This is a bandwidth increase of 1000. Notice in Fig. 9-14 that the 70.7 percent useful frequency in the open loop mode is only about 10 Hz. In the closed loop mode, when the gain has been reduced to 100 (still a

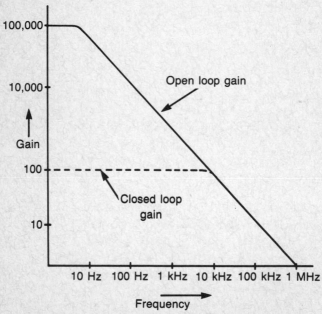

Fig. 9-14. Closed loop gain versus open loop gain of the 741 op amp.

useful amount of gain), the frequency response has been increased to 10 kHz. Again, this gain-bandwidth product is for the 741 dual supply op amp, but most other op amps are similar.

One final note before going on to several applications in which op amps are used. In the inverting amplifier of Fig. 9-15, a resistor connects the noninverting input to ground. This resistor helps minimize offset, or that condition where the output voltage should equal zero volts when both inputs are at zero volts. The value of this resistor is determined as follows:

$$R_2 = \frac{R_f R_1}{R_f + R_1}$$

In the noninverting op amp circuit of Fig. 9-16 the input resistor to the noninverting input is equal to:

$$R_2 = \frac{R_f R_1}{R_f + R_1}$$

As you can see, R_2 is the equivalent resistance of R_f and R_1 in parallel in either the inverting or the noninverting mode of operation.

OP AMP APPLICATIONS

There have been so many different ways in which op amps have been used that it would be impossible to show a circuit diagram for each of these. However, some of the more popular applications are discussed. These include the voltage follower, summing amp, differencing amp, filters, and comparator. Comparators alone would make a chapter in themselves, so many of these circuits are treated superficially just so you know what to expect at the output of the op amp for specific inputs.

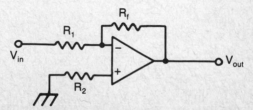

Fig. 9-15. In this inverting op amp circuit, R_2 helps to minimize offset.

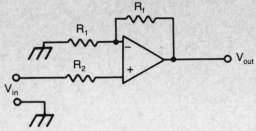

Fig. 9-16. A noninverting op amp circuit again using a resistor, R_2, to minimize offset.

The Voltage Follower

Voltage followers are circuits which have a gain of unity or slightly less and where the output signal voltage follows the phase and near amplitude of the input signal voltage. Figure 9-17 shows the op amp configuration for the voltage follower. Since the input resistance is extremely high, about 30 megohms, and the output resistance is very low, about 1 ohm, it acts very much like the emitter follower discussed earlier, making it ideal for isolation and impedance matching between stages. Since there is no R_f or R_1 to measure gain with, the equation for finding the gain of this circuit is as follows:

$$A = \frac{R_f}{R_1} + 1 = \frac{0}{0} + 1 = 1$$

The gain of this circuit configuration is 1, or unity.

The Summing Amplifier

A simple summing amplifier is shown in Fig. 9-18. This type of amplifier is also known as an analog adder. A scaling adder is shown following this discussion. In Fig. 9-18 a typical summing

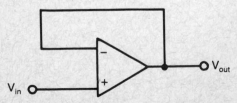

Fig. 9-17. A voltage follower (unity gain) op amp circuit is shown here.

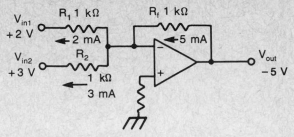

Fig. 9-18. A simple summing amplifier with two inputs.

amplifier is shown where, when R_1, R_2, and R_f are all equal in value, V_{out} equals V_{in1} plus V_{in2}. With a value of $+2$ volts at V_{in1} and a value of $+3$ volts at V_{in2}, the current through R_1 must be 2 mA while the current through R_2 must be 3 mA, in the directions shown. To pull 5 mA total through both input resistors, 5 mA must also flow through R_f since I_f equals I_{in}. But to pull 5 mA through R_f, 5 volts must be dropped across R_f, since:

$$V_{R_f} = I_f R_f = 5 \times 10^{-3} \times 1 \times 10^3 = 5 \text{ volts}$$

Now, the voltage out, V_{out}, must equal the voltage drop across R_f. The output voltage equals -5 volts. It is minus since this is an inverting amplifier. Therefore, when $R_1 = R_2 = R_f$, then $V_{out} = V_{in1} + V_{in2}$. In all cases where all input resistors are of equal value and at the same value as R_f, then $V_{out} = V_{in1} + V_{in2} + V_{in3} \cdots$, even with four or five input signal voltages.

For amplification, it's simply a matter of increasing the value of R_f by the amplification factor needed. As an example, if a gain of 10 is needed, then the feedback resistor in Fig. 9-18 should be increased from 1 k ohm to 10 k ohm.

When the input resistors are a factor of the feedback resistor, as shown in Fig. 9-19, the circuit becomes a scaling adder. In each

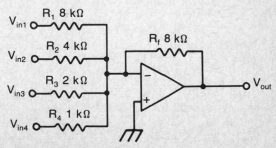

Fig. 9-19. A simple scaling adder with four inputs.

case there is an amplification of the input signal except at V_{in1}, because here the input resistor is the same value as the feedback resistor. In the scaling adder of Fig. 9-19:

$$V_{out} = -(\frac{R_f}{R_1}V_{in1} + \frac{R_f}{R_2}V_{in2} + \frac{R_f}{R_3}V_{in3} + \frac{R_f}{R_4}V_{in4})$$

Op Amp Active Filters

A filter is a device that can be used to allow a certain band of frequencies to pass while blocking frequencies above and below this band of frequencies. It can also reject a specific band of frequencies while allowing frequencies above and below this band to pass through unaffected. The first type of op amp active filter is called a bandpass filter while the latter type is called a band reject or notch filter.

To understand this better, Fig. 9-20 is an illustration, graphically, of the two different types of filter actions. The drawing on the left is of the frequency response of a bandpass filter. A specific group of frequencies is allowed to pass while frequencies above and below this band of frequencies are rejected. The illustration on the right shows the frequency response curve of a notch filter.

Figure 9-21 is a schematic of a simple bandpass filter. There is more than a resistor in the feedback loop. In this configuration, C_1 equals C_2 and R_1 equals R_2. R_1 and C_2 act as a low pass filter by providing or allowing more degenerative feedback as frequency increases while R_2 and C_1 act as a high pass filter because C_1 attenuates low frequencies. The center frequency or resonant

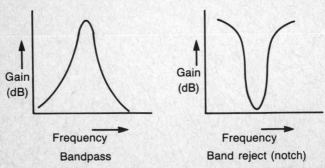

Fig. 9-20. A simple graphic representation shown here depicts the filtering action of a bandpass and a notch filter.

206

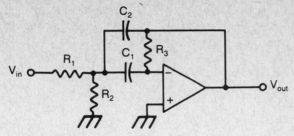

Fig. 9-21. A simple single stage op amp bandpass filter.

frequency, f_r, can be found with the equation:

$$f_r = \frac{1}{2\pi\, R_p R_3 C_1 C_2}$$

f_r = Center frequency of the pass band

$$R_p = \frac{R_1 R_2}{R_1 + R_2} = 1/2R_1 \text{ when } R_1 = R_2$$

The Q of the filter is also an important factor to consider. Q is the quality of the filter and is a measure of the ability of a filter to pass a selected frequency and reject all others. Q is a measure of the selectivity of a filter and is found with the equation:

$$Q = \frac{f_r}{BW}$$

Op Amp Differencing Amplifier

The front end or input section of an op amp is a differential amplifier, but the entire op amp itself may be used as a differencing amplifier also. This is shown in Fig. 9-22. As long as all resistors

Fig. 9-22. A typical configuration for an op amp differencing amplifier.

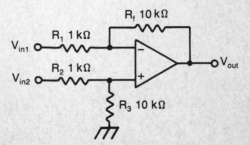

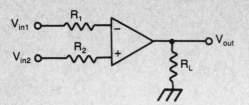

Fig. 9-23. A simple op amp voltage comparator.

are of equal value then $V_{out} = V_{in1} - V_{in2}$. If amplification of the difference signal is desired then a ratio is set up with the circuit's resistors as follows:

$$\frac{R_f}{R_1} = \frac{R_3}{R_2}$$

$$V_{out} = \frac{R_f}{R_1} (V_{in1} - V_{in2})$$

$$V_{out} = \frac{R_3}{R_2} (V_{in1} - V_{in2})$$

As an example, for a stage gain of 10 in Fig. 9-23, make R_f 10 times greater than R_1 or R_3 10 times greater than the value of R_2.

Op Amp Comparator

An op amp comparator is one of the most useful and probably one of the simplest op amp circuits. Figure 9-23 is a simple voltage comparator which compares one input signal voltage level to the other and gives an appropriate output, either saturating to a positive voltage (called railing to the supply voltage), or saturating to a negative voltage (called railing negative) at its output. Either input signal may also be compared to ground or to some fixed positive or negative voltage.

As an example, if V_{in1} is more positive than V_{in2} then the inverting side of the op amp causes the output to be driven to negative saturation. If however V_{in2} is more positive than V_{in1} then the noninverting side of the op amp causes the output to be railed to the supply voltage. This op amp can act like a switch, switching from either a positive to negative or negative to positive voltage output with specific inputs.

Chapter 10

Oscillators

Oscillators are basic amplifiers using regenerative or positive feedback to help convert a dc voltage into a pulsating dc or ac voltage. The frequency of this pulsating dc or ac voltage may be a few hertz or even a million hertz or more. Electronic oscillators are more capable of producing high frequency signals than are most mechanical oscillators. There are many different kinds of oscillators. Some provide a sine wave output, while others provide a square wave or a triangular wave shape. You'll find oscillators in almost every kind of electronic circuit. Oscillators are basically used as time keeping devices and are found in television, radio, computers, and even in many of today's electronic toys. In many instances oscillators are used to synchronize operations between electronic systems so that certain actions take place within the system at the same time or within a specific time from one system to another.

OSCILLATOR FUNDAMENTALS

A number of requirements must be satisfied before a circuit can be classified as an oscillator: the circuit must be an amplifier, it must use regenerative feedback, it must be self-excited or self-starting, and finally, to set oscillator frequency, frequency determining components must be used. When all of these aspects are in place, conditions for oscillation have been met. To better understand this criterion, Fig. 10-1 shows a simple oscillator.

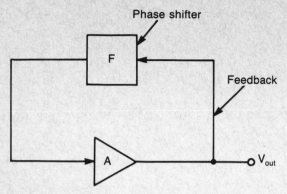

Fig. 10-1. A representation of oscillator feedback.

In this illustration, the amplifier gain is represented by A and the feedback network is labeled F. Remember, F must also shift the phase of the output if an inverting amplifier is used so that the signal fed back to the input of the amplifier is in phase with the input signal. If F equals 0.5, the signal fed back to the input of the amplifier is one half the amplifier output level. The product of amplifier gain and the feedback factor, in this case 0.5, must be equal to or greater than unity when regenerative feedback is used in order for the oscillator to be self-perpetuating and self-starting. If in the example of Fig. 10-1 the gain of this amplifier is equal to 2 or more, then oscillation occurs.

TYPES OF OSCILLATORS

Oscillators can be constructed using either discrete components, such as transistors, or with integrated circuit op amps. The IC op amp is undoubtedly the more popular of the devices used in constructing oscillators because of its simplicity and high gain and because it costs less overall than the transistor oscillator circuits. There is, however, a need to understand how oscillators are constructed using transistors as well as op amps. This section covers transistor oscillators and, where appropriate, the comparable op amp circuit is introduced as well.

Also, other oscillators are designed specifically for the uhf and microwave frequency ranges. In essence, these are not really solid state devices, but mechanical assemblies and therefore aren't discussed here. For further information consult other texts on klystrons, magnetrons, and traveling wave tubes (TWTs) (the latter sometimes is referred to as a backward wave generator or oscillator). These types of mechanical oscillators are a significant part of the hardware used in commercial and military radar applications.

Transformer Oscillators

A common-emitter amplifier has a 180° phase shift from base input to collector output and can be used as an oscillator if the collector output is fed back to the base while phase shifting the output signal by 180°. One method of accomplishing this, shown in Fig. 10-2, is referred to as a basic transformer oscillator. The biasing circuit is not shown for simplicity's sake. When the circuit is energized, Q_1 is turned on allowing current to flow from emitter to collector, up through L_1 and back to VCC. The changing current in L_1 induces a current in L_2 by transformer action. The current through L_2 being fed to the base of Q_1 is 180° out of phase from the collector of Q_1. Remember, the secondary of a transformer phase shifts the signal from its primary. This increases the base bias of Q_1 until a point is reached where the amplifier is saturated. C_1 is then charged to VCC and L_1 stops expanding causing no voltage to be induced into L_2. Q_1 now stops conducting. However, the

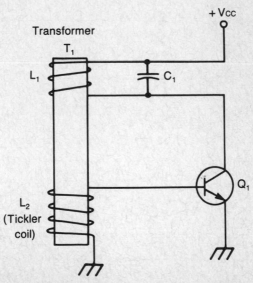

Fig. 10-2. A basic transformer oscillator utilizing a tickler coil.

field current around L_1 collapses discharging C_1 through L_1. When C_1 is completely discharged through L_1, the field in L_1 again collapses, but in the opposite direction through C_1. This second collapse drives Q_1 further into saturation. When C_1 again discharges through L_1, a positive voltage is induced into L_2, turning on Q_1. Once again C_1 charges to Vcc, Q_1 saturates again, and the cycle in the tank circuit of L_1C_1 begins again. Basically, the transistor Q_1 acts like a switch that closes to replace the energy that is lost in the tank with the circuit of L_1C_1 oscillating and providing transistor bias.

Figure 10-3 is a tuned-collector oscillator. This is basically the same circuit as that of Fig. 10-2, but with the biasing scheme shown. Also, C_1 is variable now so that the output frequency is selectable. This type of oscillator is operated class C.

A tuned-base circuit is shown in Fig. 10-4, a variation of the Armstrong oscillator, called a tuned-base oscillator because the tuned circuit is in the base circuit of Q_1. Once again, the oscillation takes place in the tank circuit L_1C_1 of the base of Q_1 inducing a voltage into L_2 that is 180° out of phase with the voltage in L_1. The charge on C_1 forward-biases Q_1 until Q_1 saturates, supplying a steady current through L_2. This prevents a voltage from being induced in L_1 allowing C_1 to discharge through L_1, starting tank circuit oscillations. When C_1 discharges, Q_1 is cut off. After a

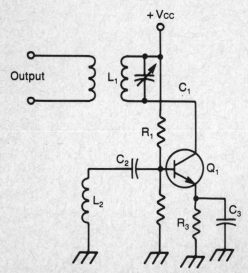

Fig. 10-3. A basic tuned-collector oscillator.

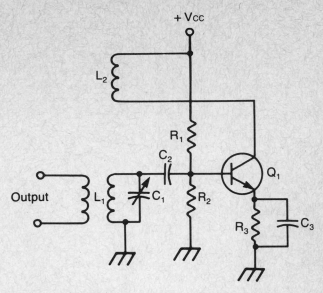

Fig. 10-4. A typical tuned-base oscillator.

complete cycle in the tank circuit, Q_1 is again forward biased and the cycle starts over again.

LC Oscillators

The LC oscillators in this section are concerned primarily with producing a sine wave output. The oscillators of Figs. 10-3 and 10-4 are LC oscillators that produce sine wave outputs by using a tank circuit in the feedback network. The problem with the tickler coil in Fig. 10-2 is that it is unstable, resulting in oscillator output frequency variations. If this tickler coil can be made part of the tuned circuit, these frequency variations can be eliminated. One of these is the series-fed Hartley oscillator shown in Fig. 10-5.

In this oscillator circuit the two coils are formed from one tapped inductor, with the tuning capacitor connected across it in parallel. When Q_1 is initially turned on, current flows up through L_{1A} and through Q_1 and R_3 to Vcc. Current is also induced into L_{1B} through mutual inductance. This results in a positive potential at the base of Q_1 by capacitor C_1 that very quickly saturates Q_1. Now that the current in L_{1A} is no longer changing, no changing current is being induced into L_{1B}. Forward bias on Q_1 decreases, decreasing current through Q_1. Next, the field on L_{1A} collapses and induces a current in L_{1B} causing a negative potential to be felt

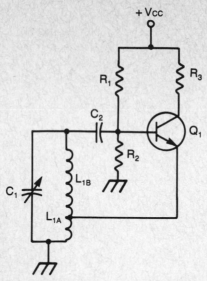

Fig. 10-5. A series-fed Hartley oscillator with tapped coil $L_{1A}L_{1B}$.

on the base of Q_1, driving it into cutoff. At the same time, C_1 charges to a negative value and when Q_1 is finally cut off completely, C_1 begins to discharge and the tank circuit begins its cycle of oscillations. When C_1 begins to charge positive on its top plate, Q_1 is forward biased, conducting current again through L_{1A}. This replaces energy lost in the tank circuit, providing the regenerative or positive feedback that is a necessary requirement for oscillations. Although emitter current flows through a portion of the tank circuit giving this oscillator its series-fed name, it's this configuration that also causes a lower circuit Q and frequency instability in this type of oscillator.

Another variation of this oscillator, seen in Fig. 10-6 is called a shunt-fed Hartley oscillator. The dc current does not pass through the tank coil, allowing better frequency stability and higher Q. The tank circuit of L_1 and C_1 establish output frequency. Bias components have been eliminated for simplicity. C_2 couples the base of Q_1 to L_1C_1. When the circuit is initially energized, Q_1 is forward biased. The initial current is felt through RFC resulting in a drop in collector voltage. This change in voltage is coupled back to L_1C_1 through C_3, giving energy to the tank circuit. A positive potential is felt at the top of L_{1B} since its lower end is at ground potential. The positive potential is fed to the base of Q_1, saturating Q_1. This causes a lack of change in collector current,

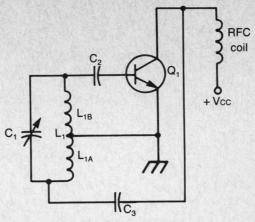

Fig. 10-6. A shunt-fed Hartley oscillator with RFC.

eliminating the ac from coupling to the bottom of L_1C_1 through C_3. The field of L_1 collapses charging the top of C_1 to a negative potential. Once the inductor field collapses completely, C_1 then discharges through the tank and reverse biases Q_1, cutting Q_1 off. Tank action completes one cycle of oscillation, charging the top of C_1 positive, turning on Q_1 and continuing the cycle.

An op amp can also perform this function as a Hartley oscillator. The identifying feature of the Hartley oscillator is the tapped inductor in the tank circuit. Figure 10-7 is an example of a Hartley oscillator where L_1C_1 form a filter that couples the selected frequency back to the amplifier input while eliminating all other

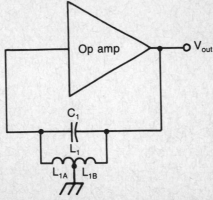

Fig. 10-7. A simplified illustration of a Hartley oscillator using an op amp as the active device.

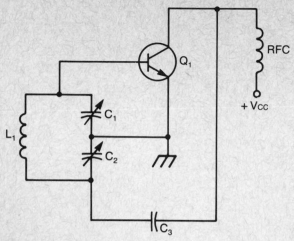

Fig. 10-8. A Colpitts oscillator like this is easily identified by the tapped capacitors.

unwanted frequencies. The oscillator frequency is then equal to the resonant frequency of the tank circuit feedback network.

The Colpitts oscillator is another tuned tank circuit oscillator that uses two capacitors in the tank circuit rather than a tapped inductor, shown in Fig. 10-8. Again, C_3 couples ac back to the tank circuit while blocking dc from the power supply. Also, RFC serves to keep ac out of the power supply while allowing dc bias for the operation of Q_1. C_1 and C_2 determine the frequency of oscillation. This circuit operates basically in the same manner as the Hartley oscillator. A Colpitts oscillator using regenerative feedback with an op amp as the amplifying device is shown in Fig. 10-9.

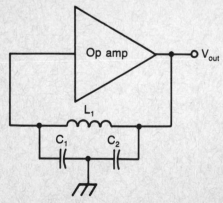

Fig. 10-9. A Colpitts oscillator using an op amp as the active device.

Again, the capacitance and inductance of the resonant tank circuit in the feedback network determine the oscillator frequency. This same feedback network also serves to couple the desired frequency of oscillation back to the input of the amplifier while rejecting all other frequencies.

The last LC oscillator to be discussed is the Clapp oscillator. Figure 10-10 is a schematic of this type of oscillator also showing the biasing scheme. The only real difference between this oscillator and the Colpitts oscillator is that C_3 has been added in series with L_1 in the tank circuit. Actually, C_3 in this oscillator determines frequency since it is relatively small in size as compared to C_1 and C_2.

Crystal Controlled Oscillators

In many cases a frequency shift of less than 1 percent is desirable in an oscillator application. LC oscillators frequently drift by 1 percent and many electronic circuits simply do not operate properly unless the frequency drift, usually caused by ambient

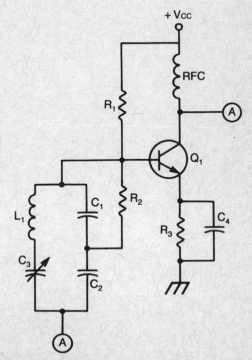

Fig. 10-10. A simplified Clapp oscillator.

temperatures and load fluctuations, is held to about 0.0001 percent. This may seem to be an impossible task, but a crystal (usually made of quartz), added to a Hartley or Colpitts oscillator can maintain the required frequency stability.

Crystals mechanically compress and stretch and resonate at a specific frequency when an ac signal is applied. This is called the piezoelectric effect and is actually the crystal resonating at its natural frequency of oscillation. The size and thickness of the crystal determine the frequency at which the crystal resonates. The thinner the crystal, the higher its natural resonating frequency. Although crystals can be sliced thin enough to vibrate at up to 50 MHz, higher frequencies are attainable by mounting the crystal in such a way so that overtones or harmonics of the original frequency are produced. This means a 10 MHz crystal can be mounted so that it vibrates at 20 MHz (the first overtone) or 30 MHz (the second overtone). Also, crystals can be made to resonate as an LC series resonant circuit resonates, with minimum impedance at the resonant frequency, or like a parallel LC resonant circuit, with maximum impedance at the resonant frequency.

Figure 10-11 is an example of a shunt-fed Hartley oscillator using a crystal in the feedback path. In this case the crystal operates in its series resonant mode and the tank circuit is tuned to the crystal frequency. The series resonant frequency of the crystal therefore

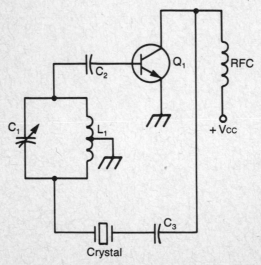

Fig. 10-11. A Hartley oscillator using a crystal to control frequency.

determines the output frequency of the oscillator. When the oscillator operates at the crystal frequency, minimum opposition to current flow exists in the feedback path due to the minimum impedance presented by the crystal. If oscillator frequency drifts, the impedance of the crystal rises sharply and reduces feedback. It forces the oscillator to return to the natural frequency of the crystal.

The Colpitts oscillator can also be better stabilized using a crystal. This is shown in Fig. 10-12. Once again, the crystal is in series with the feedback path and tuned tank circuit $L_1C_1C_2$. The crystal controls the amount of feedback in the oscillator and, like the Hartley oscillator, maintains frequency stability to a very small percentage.

Another oscillator that uses a crystal in its tank circuit in place of the coil or inductor is called the Pierce oscillator, shown in Fig. 10-13. It is similar in construction to the Colpitts oscillator. In this case the crystal is operating at its parallel resonant frequency where impedance is maximum in the equivalent crystal circuit. The ratio of C_1 to C_2 determines the feedback voltage level and therefore the voltage across the crystal, called the crystal's excitation voltage. A large voltage is developed across C_1 at resonance when the tank circuit impedance is at maximum. If the oscillator frequency drifts slightly, the impedance of the crystal quickly decreases, thus decreasing the feedback voltage. This allows the crystal to control

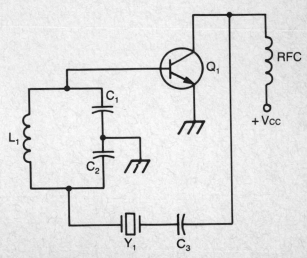

Fig. 10-12. A Colpitts oscillator using a crystal for better frequency stability.

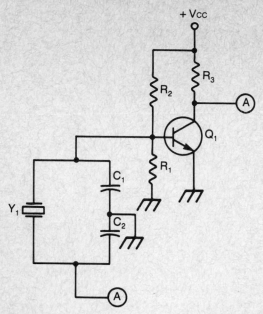

Fig. 10-13. Another type of oscillator that can use a crystal in place of tank inductance is this Pierce oscillator.

tank impedance and feedback, stabilizing the oscillator output frequency.

An op amp can also be used to construct an active Pierce crystal oscillator, as shown in Fig. 10-14. Here, the crystal is used in parallel with a capacitive voltage divider. Again, at resonance, the impedance of the feedback tank circuit is maximum with the excitation voltage of the crystal being determined by the ratio of

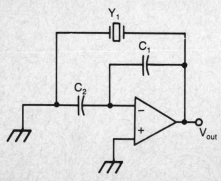

Fig. 10-14. A Pierce crystal oscillator constructed here using an op amp as the active device.

C_1 to C_2. If the output frequency of the amplifier drifts, the impedance of the crystal decreases, shunting the undesired frequencies to ground.

Another crystal oscillator is shown in Fig. 10-15. This is referred to as a Butler crystal oscillator. It uses a combination of an LC tank circuit and a crystal. The output of Q_2 is coupled back to its emitter input through Q_1 and the crystal in this feedback path. The crystal in this oscillator operates in its series-resonant mode to control the frequency stability of the oscillator.

Q_2 is forward biased when the circuit is initially energized. Biasing components are not shown, again for simplicity of explanation. As Q_2 conducts, a negative voltage appears at the bottom of the tank circuit and is thus coupled to the base of Q_1 through capacitor C_3. Once Q_2 saturates Q_1, which had been cut off, Q_2 begins to conduct. The top of R_2 now has a positive potential that is coupled to the emitter of Q_2, through the crystal. This also develops a voltage across R_1. This reverse biases the emitter-base junction of Q_2, cutting it off. This allows the collector of Q_2 to go positive, placing a positive potential on the base of Q_1 causing it to conduct harder and driving Q_2 into complete cutoff. Now the oscillation of the tank circuit of L_1C_1 reverse biases Q_1

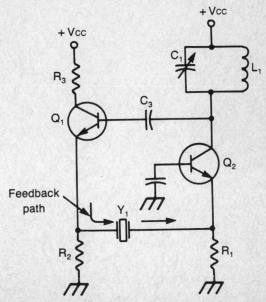

Fig. 10-15. A Butler oscillator also uses a crystal in the feedback path for better frequency stability.

221

and turns it off, causing Q_2 to be forward biased, starting the cycle all over again.

This particular oscillator offers the frequency selectivity of a crystal with the tunability of an LC tank circuit. Also, stability is increased and strain on the crystal is reduced due to a small voltage potential that is continually applied across the crystal. In effect, this particular oscillator offers a very good performance and the oscillator frequency can be changed simply by operating at an overtone of the crystal with different components in the LC tank circuit.

RC Oscillators

RC oscillators are amplifiers designed to oscillate by supplying a portion of the output signal voltage back to the amplifier input through an RC feedback network that shifts the phase of the output signal voltage by 180°. Figure 10-16 is probably the simplest of the oscillators studied thus far.

Here each RC pair (R_1C_1, R_2C_2, R_3C_3) shifts the output of the common-emitter transistor amplifier by 60°. The combined phase shift therefore, is 180° since three RC networks are used. Considerable power loss is usually found across RC networks; therefore, the gain of this common-emitter amplifier oscillator should be above 40. In this oscillator, frequency of operation can be found from the formula:

$$f_o = \frac{1}{2 \pi RC}$$

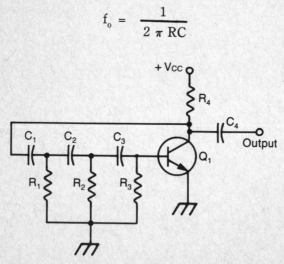

Fig. 10-16. A simple RC or phase shift oscillator.

when $R_1C_1 = R_2C_2 = R_3C_3$

f_o = Frequency of operation
R = Value of one resistor
C = Value of one capacitor

Figure 10-17 is an example of an RC phase shifting oscillator that uses an operational amplifier instead of a discrete transistor. Here f_o is found with the following equation:

$$f_o = \frac{1}{2 \pi RC}$$

when $R_1C_1 = R_2C_2 = R_3C_3$

f_o = Operating frequency
R = Value of one resistor
C = Value of one capacitor

Another type of RC oscillator is the Wien-bridge oscillator. This oscillator also uses an RC network in the feedback loop, but the purpose of the RC network is not to phase invert the output of the amplifier, but to actually select the desired operating frequency of the oscillator.

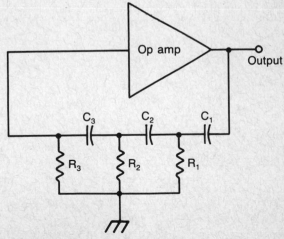

Fig. 10-17. An RC phase shifting oscillator using an op amp as the active device.

223

A discrete circuit requires a Darlington configuration at its input with a common-emitter output and a number of other components for biasing and for the bridge network as well. With the popularity of the op amp and its easier explanation of operation, the active Wien-bridge oscillator shown in Fig. 10-18 uses an op amp for the active device.

Notice the bridge network at the input to the op amp. The top of the bridge is composed of R_1C_1 and R_2C_2, a frequency sensitive lead-lag network that provides positive or regenerative feedback to the noninverting input of the op amp. The bottom portion of the network is made up of R_3 and R_4. Since no phase shifting occurs to the amplifier output passing through this half of the Wien-bridge, the signal from this side of the network allows negative or degenerative feedback to be placed at the inverting input of the op amp. For oscillations to occur, positive feedback is made larger than negative feedback by proper component selection. If frequency decreases, the reactance of C_1 increases developing less voltage across R_2C_2, reducing positive feedback. If frequency increases, the reactance of C_2 decreases, shunting positive feedback to ground. Positive feedback, therefore, will only be large enough to sustain oscillations over a very narrow range of frequencies. Operating frequency can be found from the following equation:

$$f_r = \frac{1}{2 \pi RC}$$

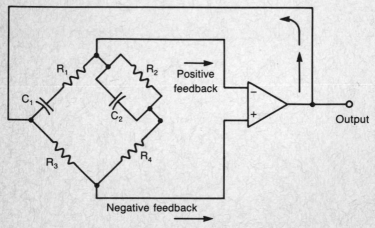

Fig. 10-18. An active Wien-bridge oscillator using an op amp.

Unijunction Oscillators

Up to this point the oscillators presented have been those that produce a sinusoidal waveform at their output. Some oscillators produce other than a sinusoidal waveform. These oscillators produce square waveshapes, triangular waveshapes, pulsed outputs, or a combination of all three. Usually oscillators that produce more than one type of output are called generators.

Oscillators that use active devices have not yet been discussed. These are the unijunction transistor (UJT) and the programmable unijunction transistor (PUT). Actually these devices fall under the category of thyristors, but are still used as the active device in some RC oscillator circuits and are presented here.

To further understand the unijunction RC oscillator it is necessary to describe the operating characteristics of the unijunction transistor. Figure 10-19 is an example of the basic construction of a unijunction transistor.

This device has a single junction made by diffusing a pellet of P-type material into an N-type semiconductor. Base 1 and base 2 are simply each end of the semiconductor. In this structure, the pellet, which becomes the emitter, is located closer to base 2 than to base 1. The resistance between base 1 and base 2 is usually quite high, approximately 10 k ohms, because the N-type material is only lightly doped and contains very few majority carriers.

Figure 10-20 is an equivalent electrical circuit of the unijunction

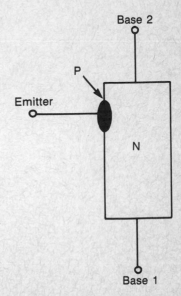

Fig. 10-19. A simplified illustration showing the construction of a unijunction transistor.

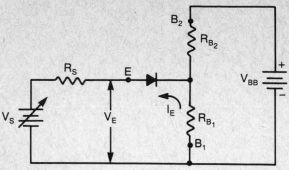

Fig. 10-20. An equivalent electrical circuit of a unijunction transistor with proper biasing.

transistor shown with proper biasing. R_s is used to limit the current through the emitter to a safe value. An important characteristic of the unijunction transistor specified by the manufacturer is called the intrinsic standoff ratio and occurs when V_S is not large enough to forward bias the diode in Fig. 10-20. When that occurs the combined voltage drop across R_{B_2} and R_{B_1} is equivalent to the voltage V_{BB} and the intrinsic standoff ratio is equivalent to the voltage drop across R_{B_1} and V_{BB} and is therefore found by:

$$\eta = \frac{V_{B_1}}{V_{BB}}$$

η = Intrinsic standoff ratio
V_{B_1} = Voltage drop across R_{B_1}
V_{BB} = Base bias voltage

Since the voltage drops are proportional to the value of R_{B_1} and R_{B_2}, then:

$$\eta = \frac{R_{B_1}}{R_{B_1} + R_{B_2}}$$

In effect, the intrinsic standoff ratio is determined by the physical construction of the device, since R_{B_1} and R_{B_2} are internal resistances, and not by controlling V_{BB} or V_S. If the intrinsic standoff ratio is specified, V_{B_1} can be found with:

$$V_{B_1} = V_{BB}$$

Also, V_{B_1}, since it is larger than V_E, keeps the diode reverse biased. It is only when V_E is larger than V_{B_1} by at least 0.7 volts (the voltage required to cause the diode to conduct current) that the UJT functions differently than before. The value required to turn on D_1 is called the peak voltage, V_P, and is found with:

$$V_P = V_{BB} + V_F.$$

V_P = Peak voltage
η = Intrinsic standoff ratio
V_{BB} = Base bias voltage
V_F = Voltage forward required to turn on D_1

Once V_P is reached and the PN junction is forward biased, the N-type semiconductor begins to allow many holes to flow from B_2 to B_1. This decreases the resistance of R_{B_1} and increases I_E (allows more electrons to flow from B_1 to the emitter). This decrease in R_{B_1} causes I_E to increase while V_E also decreases, resulting in a negative resistance characteristic of the unijunction transistor once V_P is reached. This V-I characteristic curve that takes place between E and B_1 is shown in Fig. 10-21. Here, once V_P is reached, V_E decreases in spite of an increase in I_E. Once the valley voltage, V_V, is reached, V_E begins to increase when I_E increases.

Figure 10-22 is an oscillator circuit using a unijunction transistor (notice the symbol), as the active device. The RC time constant of R_1C_1 is used to determine oscillator frequency. This circuit is

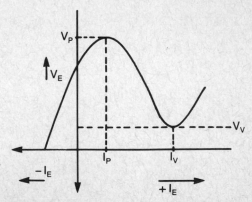

Fig. 10-21. The V-I curve of a typical unijunction transistor.

227

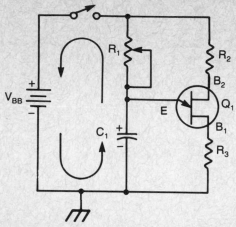

Fig. 10-22. A relaxation oscillator using a unijunction transistor.

called a relaxation oscillator and can actually generate three different waveforms. When V_{CC} is first applied, C_1 charges with a ground potential at its bottom half and a positive potential at its top half. When the value of V_P is reached, resistance between E and B_1 decreases. C_1 then discharges through the UJT and R_3. Q_1 then turns off when the voltage across C_1 discharges to the value of V_V. This allows C_1 to charge up again, repeating the cycle. V_P slowly increases in value while V_V drops rapidly in value producing a sawtooth waveform from the emitter of Q_1. The quick discharge of C_1 through Q_1 and R_3 produces pulses on B_1. These pulses are especially useful as triggering pulses in SCR applications. It is the intrinsic standoff ratio that determines the device triggering voltage, or V_P.

Another device used as an active device in RC oscillator circuits is the programmable unijunction transistor, or PUT. It is similar to its cousin the UJT, but is considerably more temperature stable, and allows the intrinsic standoff ratio to be determined by values of external resistors rather than the internal physical structure of the base resistances which are determined during manufacture. Figure 10-23 is an example of a PUT used as an active device in a relaxation oscillator. Notice the symbol used for the PUT. It has three terminals: an anode, cathode, and gate.

Resistor R_2 has a low value and is used to develop the output voltage pulse of the oscillator. R_4 develops gate voltage. When the circuit is energized, C_1 charges up from ground through R_1. When C_1 charges up to the value of V_P, the PUT turns on and C_1

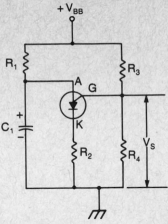

Fig. 10-23. A relaxation oscillator using a PUT as the active device.

discharges through the PUT and R_2. Again, as with the UJT, when the value of voltage on C_1 drops to the valley voltage, V_V, the PUT turns off and the cycle begins again. Again, C_1 develops a sawtooth waveform while R_2 develops the output pulses. What is unique in this circuit is that, although frequency can be varied by varying R_1 or C_1, it can also be varied by changing the ratio of R_3 to R_4. This ratio controls the gate voltage and thus controls the value of peak voltage, V_P. To increase frequency, R_3 is increased while R_4 is held constant. To decrease frequency, R_3 is held constant while R_4 is made larger in value.

In Fig. 10-23 the following equation holds true as it did for the UJT:

$$\text{Standoff ratio} = \frac{R_4}{R_3 + R_4}$$

Although the ratio of R_3 to R_4 determines V_P, I_P and I_V for any given value of V_S are actually a function of the values of each individual resistor. Also, as gate resistance R_G increases for any given value of V_S, I_P and I_V decrease. R_G can be found with:

$$R_G = \frac{R_3 R_4}{R_3 + R_4}$$

Figures 10-24 through 10-26 are examples of data sheets given by the manufacturer for some PUTs.

Planar, TO-18 Hermetic

ABSOLUTE MAXIMUM RATINGS

Anode-to-Cathode Forward Voltage, V_{AK}	40V
Anode-to-Cathode Reverse Voltage, V_{AKR}	40V
Gate-to-Cathode Forward Voltage, V_{GK}	40V
Gate-to-Anode Reverse Voltage, V_{GAR}	40V
Gate-to-Cathode Reverse Voltage, V_{GKR}	5V
Peak Recurrent Forward Current	
10 μs 1% Duty Cycle	8A
100 μs 1% Duty Cycle	5A
Power Dissipation	
25°C Ambient	400mW
Derating Factor	3.2mW/°C
Storage Temperature Range	−55°C to +150°C
Operating Temperature Range	−55°C to +150°C

DESCRIPTION

The Unitrode hermetically sealed TO-18 metal can series of programmable unijunction transistors feature blocking voltages to 100V, the highest available to designers. These PUTs are functionally equivalent to standard unijunction transistors, with the added advantages of programming versatility. External resistors can be added to program η, R_{BB}, I_p and I_v, depending upon your design requirements. All units are fully planar passivated. This series features a hermetically sealed TO-18 package for optimum reliability in all environmental conditions. Applications include pulse and timing circuits, SCR trigger circuits, relaxation oscillators, and sensing circuits. For further application information see Unitrode's Application Note 'J-66.

TO-18

FEATURES

- Voltage Ratings: to 100V
- Maximum Peak Current: 150nA
- Valley Current: as low as 25 μA
- Low Forward Voltage Drop
- Nano-Amp Leakage
- Hermetically Sealed TO-18 Metal Can

⊔ **UNITRODE**

MECHANICAL SPECIFICATIONS

U13T1–U13T2

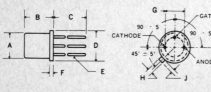

GATE CONNECTED TO CASE

	INCHES	MILLIMETERS
A	.178–.195 DIA.	4.52–4.95 DIA.
B	.170–.210	4.31–5.33
C	.5 MIN.	12.70 MIN.
D	.209–.230 DIA.	5.31–5.84 DIA.
E	.017 ± $^{.002}_{.001}$ DIA.	.432 ± $^{.051}_{.025}$
F	.020 MAX.	.508 MAX.
G	.100±.010 DIA.	2.54±.254 DIA.
H	.041±.005	1.04±.127
J	.028–.048	.711–1.22

Fig. 10-24. A typical section of a PUT data sheet (courtesy Unitrode Corporation).

ELECTRICAL SPECIFICATIONS (at 25°C unless noted)

Test	Symbol	Fig.	U13T1 Min.	U13T1 Max.	U13T2 Min.	U13T2 Max.	Units
Peak Current	I_P	1	—	5	—	1.0	μA
			—	2	—	0.15	μA
Valley Current	I_V	1	70	—	25	—	μA
			—	50	—	25	μA
Offset Voltage	V_T	1	0.2	0.6	0.2	0.6	V
			0.2	1.6	0.2	0.6	V
Gate-to-Anode Leakage	I_{GAO}	2	—	10	—	10	nA
			—	100	—	100	nA
Gate-to-Cathode Leakage	I_{GKS}	3	—	100	—	100	nA
Forward Voltage	V_F	4	—	1.5	—	1.5	V
Pulse Output Voltage	V_o	5	6	—	6	—	V
Pulse Output Rate of Rise	t_r	5	—	80	—	80	nS

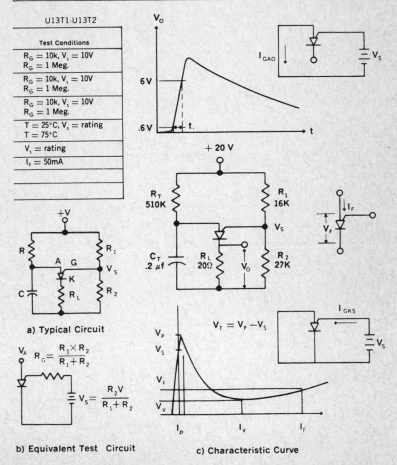

U13T1-U13T2

Test Conditions

$R_G = 10k$, $V_s = 10V$ $R_G = 1$ Meg.
$R_G = 10k$, $V_s = 10V$ $R_G = 1$ Meg.
$R_G = 10k$, $V_s = 10V$ $R_G = 1$ Meg.
$T = 25°C$, $V_s = $ rating $T = 75°C$
$V_s = $ rating
$I_F = 50mA$

a) Typical Circuit

$$R_G = \frac{R_1 \times R_2}{R_1 + R_2}$$

$$V_s = \frac{R_2 V}{R_1 + R_2}$$

b) Equivalent Test Circuit

$$V_T = V_P - V_S$$

c) Characteristic Curve

Fig. 10-25. Additional information on the PUT of Fig. 10-24 is seen here (courtesy Unitrode Corporation).

TUNABLE FREQUENCY OSCILLATORS

Variable oscillator circuits which include active elements for discharging the timing capacitor C_T are shown in Fig. 7. A second method is given as in Fig. 8.

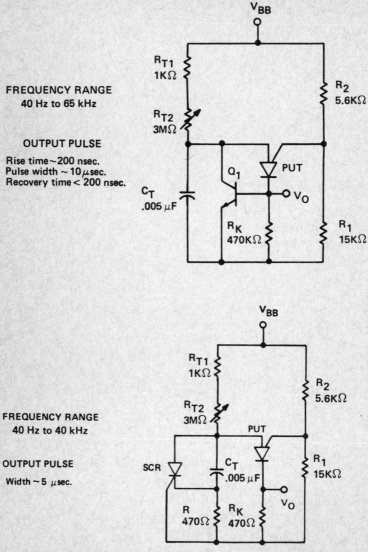

FREQUENCY RANGE
40 Hz to 65 kHz

OUTPUT PULSE

Rise time ~ 200 nsec.
Pulse width ~ 10 μsec.
Recovery time < 200 nsec.

FREQUENCY RANGE
40 Hz to 40 kHz

OUTPUT PULSE

Width ~ 5 μsec.

Fig. 10-26. Some databooks have application notes, like this one, on the PUT and practical applications as well (courtesy Unitrode Corporation).

Chapter 11

Pulse Circuits

The oscillators studied thus far generally produce a sine wave output. An oscillator is really a signal generator that can produce not only sine wave outputs, but other types of outputs as well. A sawtooth waveform and a type of spike output used for triggering purposes can also be supplied. These nonsinusoidal outputs can be measured not only in terms of amplitude, but also in terms of frequency duration and width. These types of nonsinusoidal waveforms can also be classified as pulses since they have an obvious rise time and sometimes different fall times, normally not the case with a consistant cycle of signal voltage frequency, a characteristic of the simple sine wave. Some authors include nonsinusoidal patterns as oscillators, but I prefer to view these types of oscillators more as pulse generators. Therefore this chapter includes, in a general sense, nonsinusoidal oscillators, or pulse circuits, since they are closely interrelated. Types of waveforms, the circuits used to produce those waveforms, and terminology are discussed.

NONSINUSOIDAL WAVEFORMS

A pulse circuit produces a pulse waveform. This may sound obvious, but some oscillators also produce pulses and are not considered pulse circuits but merely oscillators. These pulses are nonsinusoidal in nature and therefore have characteristics that are different from that of the simple sine wave.

Two methods are used to analyze and display waveforms. The first is called time-domain analysis—the waveform is displayed according to its amplitude over a specific period of time. A waveform appearing on an oscilloscope display screen is an example of time-domain analysis. The amplitude or domain of the waveform is displayed over a period of time. Figure 11-1 is an example of several different types of waveforms, their amplitudes, and a period of time T_2, over which these waveforms are displayed for analysis. The time periods are used so that you can see what the waveform is doing at a particular time after pulsing or oscillations begin. In some circuits where timing is critical, time-domain analysis is used to compare two or more waveforms to each other so that you can tell at a glance if specific points in a circuit are functioning at the correct time in relationship to each other or to other points in the circuit. There are special pieces of electronics test equipment, called logic analyzers, that display several waveforms at once so that time-domain analysis can be completed more easily.

The second method of analyzing circuits is through the use of frequency domain analysis. This type of analysis is based on the fact that any repeating, or periodic, waveform is actually composed of a number of sine waves. In fact, a square wave is made up of odd harmonics of a single sine wave. As an example, a 1 kHz square

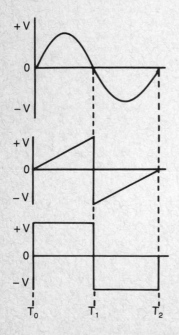

Fig. 11-1. Several different types of waveforms, two of which are nonsinusoidal.

234

wave is made up of the fundamental 1 kHz sine wave, 3 kHz third harmonic of the 1 kHz sine wave, 5 kHz fifth harmonic of the 1 kHz sine wave, and continuing, in theory at least, ad infinitum. A square wave is made up of the fundamental frequency of the square wave and all of the odd harmonics of the original sine wave which is at the same frequency as the final square wave.

Another type of waveform made by adding both odd and even harmonics of a specific frequency sine wave is called a sawtooth wave. You are familiar with this type of waveform from the previous chapter. By superimposing waves of specific frequency characteristics, phase, and amplitude over one another, almost any type of waveform can be produced. To display these waveforms in the frequency domain requires the use of a display device called a spectrum analyzer. Figure 11-2 shows how a square wave can be displayed on a spectrum analyzer using frequency-domain analysis. Figure 11-3 shows the display of a sawtooth waveform using the same spectrum analyzer. The pulses that are shown represent the harmonic content of the waveform.

Along with analysis of certain waveforms using either of these two methods, it is essential to understand the terminology used to deal with pulse circuits. Figure 11-4 is an example of a periodic wave and how the period of the wave is measured, along with the pulse width of the pulse. To determine the rate or frequency at which the pulse occurs (how many times in one second that the waveform completes one cycle), the following equation is used:

$$f = \frac{1}{period}$$

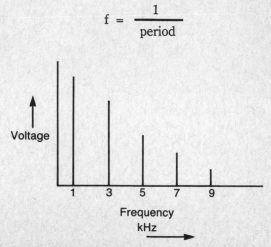

Fig. 11-2. Analysis of a square wave using frequency domain.

235

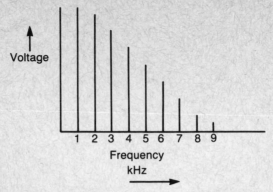

Fig. 11-3. Analysis of a sawtooth waveform using frequency domain.

In Fig. 11-4, if the period is 1 millisecond (0.001), the frequency can be found thus:

$$f = \frac{1}{period}$$

$$f = \frac{1}{0.001} = 1000Hz = 1kHz$$

Frequency is measured in hertz. The relationship between pulse width and period is called duty cycle and can be expressed as a ratio:

$$duty\ cycle = \frac{pulse\ width}{period}$$

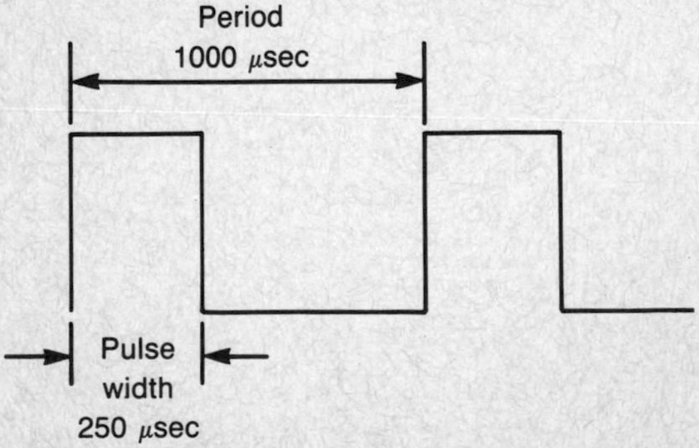

Period
1000 μsec

Pulse
width
250 μsec

Fig. 11-4. A square wave showing how pulse width and the period of the wave are measured.

236

If the pulse width in Fig. 11-4 is equal to 250 microseconds and the period is 1 millisecond, or 1000 microseconds, then the duty cycle is:

$$\text{duty cycle} = \frac{250}{1000} = 0.25 = 25\%$$

The answer in this last equation is given as a percentage because the duty cycle represents the percentage of one complete cycle in which the pulse exists.

Figure 11-5 is an example of an expanded pulse. This single pulse in itself has its own terminology in describing its make-up or characteristics. The first of these is the rise time of the pulse: the time that is required for the leading edge of the pulse to rise from 10 percent to 90 percent of its maximum amplitude. In the same respect the fall time is the time that is required for the trailing edge of the pulse to decay or fall from 90 percent to 10 percent of its maximum amplitude.

Figure 11-6 is an example of a little closer look at the pulse of Fig. 11-5. Here, several imperfections or distortions in the waveform are evident. These are called overshoot, undershoot, ringing, and settling time. Overshoot occurs when the leading edge of the pulse rises above, or overshoots, its allowable maximum amplitude. A result of this is ringing which is damped oscillations following the overshoot. In essence, the maximum amplitude is trying to return or level out to a linear value. In some high frequency pulse circuits ringing continues across the entire top of the square

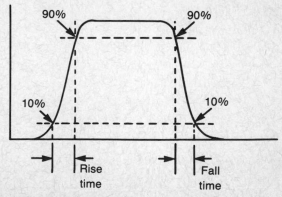

Fig. 11-5. A typical square wave showing rise and fall times.

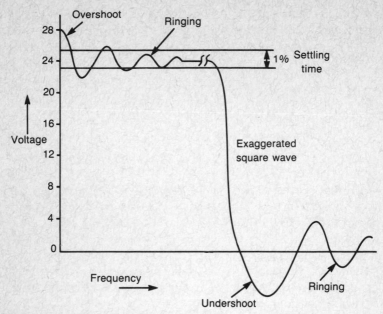

Fig. 11-6. A closer look at Fig. 11-5 with some of its imperfections.

wave. An oscillator known as a ringing oscillator utilizes this concept, but it's not very desirable in most pulse circuits and is normally limited to only a few cycles.

Settling time is usually the amount of time required for the ringing to be reduced to a maximum of 1 percent of the amplitude of the periodic pulse. Undershoot is also accompanied by ringing and this is shown in Fig. 11-6. Again, settling time can also be specified here.

WAVESHAPING

Waveshaping is an important function in some electronic circuits. It is necessary, in some cases, to change a sine wave to a square wave, or a rectangular waveform to a pulse waveform. The circuits to be discussed in this section change the shape of the input waveform so that it can be used for a particular application in an electronic circuit.

The Differentiator

There are times when it is necessary to know how a variable in an electronic circuit changes with respect to a specific period of time. An example of this is the equation used to find the counter cemf that is generated in a coil having an inductance L:

$$v = -L \frac{di}{dt}$$

v = the cemf
$-L$ = the inductance of the coil
$\frac{di}{dt}$ = the rate of change of current with respect to time

The last part of this equation involves finding the rate of change of one variable, in this case current, with respect to another, in this example, time. This is also known as differentiating and is used in some electronic circuits.

In practical terms, a differentiator is a circuit that is used to change a square wave input into an output consisting of a series of spikes. This is differentiating because these output spikes are actually proportional to the rate of change of the input signal. In this respect the RC time constant of the differentiator circuit of Fig. 11-7 is extremely important. In this differentiator circuit, the RC time constant must be very short as compared to the period of the input square wave. You can follow how the circuit of Fig. 11-7 differentiates by looking at the waveforms of Fig. 11-8 and reading the following circuit description.

At T_0, the rate of change or frequency at which the input voltage changes from zero volts to its maximum positive voltage is very high. This causes the capacitive reactance of C_1 to be very low compared to the resistance of R_1. In effect then, almost the full voltage drop is developed across R_1 instead of C_1 at time T_0. C_1 begins to charge causing current flow through R_1. Eventually C_1 charges completely, its charging time determined by the RC time

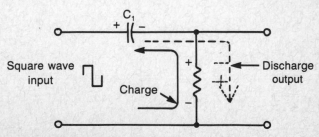

Fig. 11-7. A simple differentiator circuit showing charging and discharging paths.

239

constant of the circuit, and the voltage previously developed across R_1 drops quickly to zero volts at time T_1.

At time T_1, the input voltage pulse changes from its maximum positive amplitude level to zero volts. Now C_1 discharges through R_1 in the opposite direction in which it is charged, developing a voltage potential across R_1 that is also in the opposite direction. The output waveform therefore, also goes in the opposite or negative direction as shown in Fig. 11-8 at time T_1. When the capacitor is completely discharged, the output voltage returns to zero volts at time T_2. An interesting point here is that at any given time both the voltage across the resistor plus the voltage across the capacitor must be equal to the value of the input voltage. Notice that when R_1 was dropping the complete input voltage, C_1 had not yet begun to charge up. But as C_1 began to charge, the voltage across R_1 began to decrease until it reached zero volts. At that time C_1 was then completely charged. As the RC time constant is made larger, the output waveform will more closely resemble the input waveform.

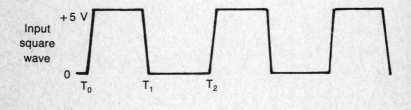

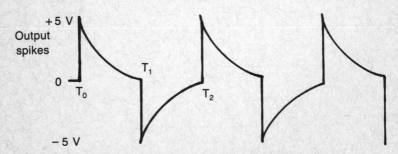

Fig. 11-8. Input and output waveforms of a differentiator circuit.

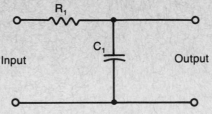

Fig. 11-9. A simplified schematic of an integrator.

The Integrator

If the output waveform were taken across the capacitor instead of the resistor, as in Fig. 11-9 the waveform of Fig. 11-10 would result. When the input waveform at time T_0 changes from zero volts to its maximum positive voltage, C_1 charges up until at time T_1 it is fully charged. This exponential charging rate should be familiar to you. At time T_1 the input voltage now drops back to zero volts, the capacitor discharges, and the resulting output is again shown in Fig. 11-10.

The integrator can also be used to develop a triangular waveform, to be discussed later, using an operational amplifier.

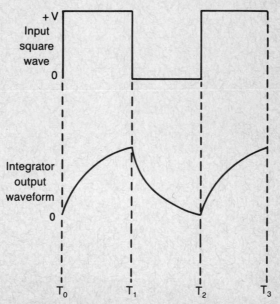

Fig. 11-10. The input and output waveforms of an integrator.

Both the differentiator and the integrator are circuits that change an input square wave to a particular kind of output waveform. If a sine wave is used instead of a square wave in either of these circuits, the output is also a sine wave. Only complex waveforms can be differentiated or integrated.

CLIPPING CIRCUITS

A clipping circuit, or peak clipper, is, according to IEEE definition, "a device that automatically limits the instantaneous value of the output to a predetermined maximum value." In effect, the clipper clips off the peak portion of the input signal and retains it for further processing. A limiter is also a form of a clipper but the top or bottom portion of the input signal is kept to a specified limit in the output. Figure 11-11 is an example of how a clipper operates, and Fig. 11-12 is an example of how a limiter prevents the output waveform from exceeding specified positive and negative values. There are, of course, limiters that limit either the top or bottom portion only of the waveform. The transfer line of Fig. 11-12 is simply a specified limiting line drawn across the input waveform as that waveform changes or transfers to its output waveform. In effect, it allows you to see what limiting action takes place in the circuit from input to output.

Clippers and limiters are very similar, and when a clipper is used to prevent a voltage from exceeding a certain level, as it often is, it is referred to as a limiter. The circuits to be discussed in this section are those of the diode and transistor clipping circuits.

Diode Clippers

Diodes make excellent clippers because they pass current in one direction only. This means that they can clip off any voltage above or below a certain reference level. A half wave rectifier is one form of diode clipper shown in Fig. 11-13. Here, the diode is in series with the output load resistor R_L.

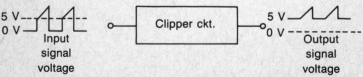

Fig. 11-11. A basic clipping circuit operation showing input and output signal voltages.

242

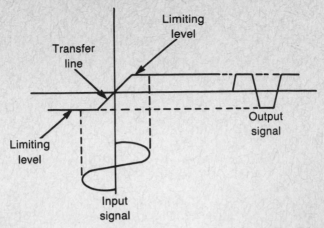

Fig. 11-12. A graphic representation of how a limiter limits the input signal waveform.

The diode in this circuit only conducts on the negative alternation of the input signal because the cathode must be negative with respect to the anode to be forward biased. Once D_1 conducts, it acts like a closed switch. This forces current through R_L, developing the output waveform shown. The diode could also be turned around resulting in a waveform in which the bottom half of the input signal voltage would be clipped off. In reality, the diode drops about 0.7 volts which means the voltage drop across R_L is actually 0.7 volts less than shown, but for simplicity, this small voltage drop is ignored in these circuit descriptions. Since the diode is in series with the output, this circuit is called a series clipper.

In Fig. 11-13, the clipping level is at zero volts. A zero volt clipping level may not be the desired level in some circuit applications, so a series clipper can be biased so that the clipping

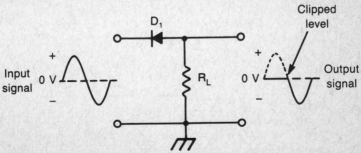

Fig. 11-13. A simple series clipper with input and output waveforms.

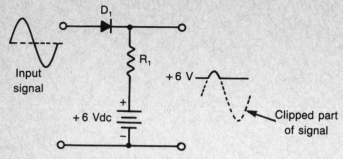

Fig. 11-14. A basic series clipper with biasing.

level is determined by the biasing source. Figure 11-14 is an example of a biased series clipper showing input and output waveforms. The biasing source holds the cathode of D_1 at +6 volts. The diode does not conduct until the input signal is 0.7 volts more positive than the biasing voltage, or +6.7 volts. The output remains at +6 volts except for that portion of the positive input signal that exceeds +6 volts. All of the negative portion of the input signal waveform has been clipped along with part of the positive input waveform. Again, the diode and biasing voltage source can be turned around resulting in all of the positive portion and some of the negative portion of the input waveform being clipped. In both cases, the reference level has been changed from zero to a positive or negative level.

Another type of clipper is shown in Fig. 11-15. This is called a shunt clipper because the diode, D_1, is in parallel, or shunted, across the output. On the positive alternation of the input signal voltage, D_1 conducts. The only output voltage developed across D_1 when it conducts is the voltage drop of 0.7 volts across its PN junction. The output allows 0.7 volts of the positive input signal voltage and all of the negative input signal voltage to pass to the output as shown in Fig. 11-15. If the negative portion and 0.7 volts

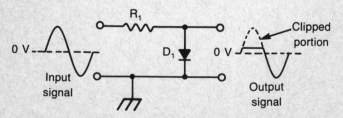

Fig. 11-15. A basic shunt clipper with input and output waveforms.

244

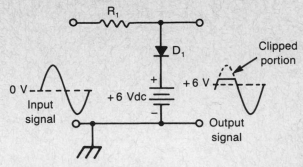

Fig. 11-16. A shunt clipper with biasing.

of the positive portion is to be clipped, D_1 would have to be reversed.

As with the biased series clipper, the shunt clipper can also be biased, shown in Fig. 11-16. Here, the output reference voltage is changed from zero volts to $+6$ volts. The cathode of D_1 is held to $+6$ volts by the biasing voltage source. For D_1 to conduct, its anode must be 0.7 volts more positive than its cathode, or in this case, $+6.7$ volts. When D_1 conducts, the output voltage is clamped to $+6.7$ volts which is the $+6$ volts of the biasing voltage source plus the 0.7 volt drop of the diode.

A negative biased shunt clipper can also be constructed simply by reversing the diode and biasing voltage. In that circuit all of the negative input signal voltage is clipped except -0.7 volts.

There are clippers that limit both the positive and negative portions of the input signal voltage. In this arrangement the circuit of Fig. 11-17 is actually referred to as a limiter because the output signal voltage is a linear representation of the input signal voltage, but its positive and negative amplitudes have been limited to a predetermined level. In this circuit the output is taken across the parallel combination of D_1 and D_2 held at 6 volts. When the

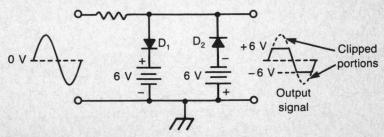

Fig. 11-17. A circuit that clips on both the negative and positive portions of the input signal.

negative portion of the input signal voltage reaches -6.7 volts, D_2 conducts and limits or changes the negative portion of the output signal voltage to this level. D_1 conducts when the positive portion of the input signal voltage reaches $+6.7$ volts, again clamping or limiting the positive portion of the output signal voltage to this level. If neither of the input voltage levels reaches these plus and minus values, the input sine wave is simply passed to the output. Incidentally, other types of waves such as square waves may also be used as an input signal voltage source.

The Transistor Clipper

Transistors can also be used as clippers or limiters. If they amplify the clipped output they are called active clippers or active limiters. In the transistor circuit of Fig. 11-18, Q_1 is used merely to change the input sine wave to a rectangular wave of the same frequency as shown in Fig. 11-19. These waveforms help to explain the operation of the transistor clipper.

Initially, when the power is applied, the transistor is cut off and the output voltage is $+$ Vcc. At time T_0, when the input sine wave reaches $+0.7$ volts, the transistor begins to conduct. As the input goes more positive the transistor conducts harder until at time T_1, Q_1 saturates and the output then falls to $+V_{CE(sat)}$. Once the positive input signal voltage falls to below the saturation level at time T_2, the transistor begins to conduct again in its linear region from time T_2 to T_3. At time T_3, the transistor again is in cutoff and the output voltage across Q_1 also becomes $+$ Vcc.

CLAMPERS

In some amplifying stages of electronic circuits the dc reference

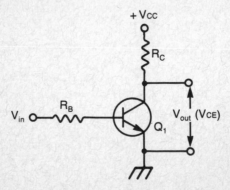

Fig. 11-18. A transistor clipper used to change a sine wave into a square wave.

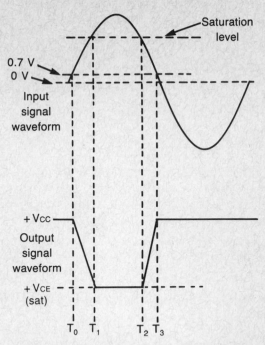

Fig. 11-19. Input and output waveforms of the clipper of Fig. 11-18.

level can be changed or lost completely. A clamping circuit can restore this dc reference level and is therefore sometimes called a dc restorer. Clamping circuits change the dc reference voltage of a waveform by clamping the top or bottom of the input signal waveform to a specified dc voltage. Clamping may be to zero volts, a positive dc voltage, or a negative dc voltage. Clampers do not, however, change the shape of the waveform.

A Simple Dc Clamper

The simplest clamping circuit is the dc clamping circuit of Fig. 11-20. This circuit is used to clamp the bottom of the input square wave to zero volts. It must be assumed here that the time constant of the discharge circuit R_1C_1 is so long that C_1 does not have a chance to discharge between half-cycles of the input signal voltage waveform and therefore, after a few input cycles, C_1 charges up to the peak voltage of the positive half-cycle of $+5$ volts.

The circuit can now be redrawn to look like the circuit of Fig. 11-21 where the charged capacitor C_1 is represented as a battery in series with the input signal voltage. At time T_0 the positive

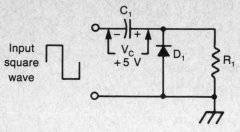

Fig. 11-20. A simple dc clamping circuit.

portion of the input signal voltage, + 5 V, adds to the + 5 volts of the capacitor and the output signal voltage is + 10 volts. At time T_1 the negative portion of the input waveform goes to − 5 volts. The difference between that − 5 volts and the + 5 volts across the capacitor equals zero volts, the output voltage of the circuit. By adding a forward biasing voltage source in series with the diode, as was done with the clipper, the bottom portion of the output voltage signal can be referenced to a positive voltage value rather than zero volts, by the amount of voltage level of the biasing source. And by reversing the diode and using forward bias, the top portion of the output voltage signal can be referenced to a negative voltage value rather than zero volts, again by the amount of the voltage level equal to the biasing source.

RECTANGULAR WAVEFORM GENERATORS

Entire books have been written about waveform generators and since space is a factor here, only the more popular types of waveform generators are discussed.

Some of the rectangular waveform generators are actually free running, requiring no input signal to produce an output signal, while others produce an output waveform only when triggered by an input signal waveform. Many of these generators are used for timing

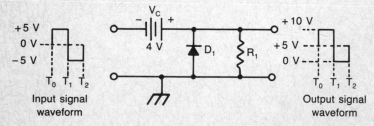

Fig. 11-21. Capacitor C_1 of Fig. 11-20 is represented here by a battery with input and output waveforms shown.

248

purposes or to produce other waveforms such as the sawtooth and triangle waveforms.

The Astable Multivibrator

This generator is sometimes referred to as a free-running multivibrator because it requires no input signal in order to produce an output signal. Astable multivibrators can be made from a circuit using two transistors or from a circuit using an operational amplifier. Figure 11-22 is an astable multivibrator circuit, sometimes referred to as a collector-coupled multivibrator because the output of each collector is coupled back to the input of the other transistor.

Basically, one transistor is conducting while the other transistor is cut off. This condition reverses on each transistor, or oscillates from one to the other. Frequency is determined by $R_2R_3C_1$ and C_2. Figure 11-23 is the result of the oscillations. The circuit operation can be analyzed by viewing both Fig. 11-22 and Fig. 11-23.

When the circuit is energized, one transistor begins to conduct more quickly than the other. Assume Q_2 begins to conduct first. This arbitrary start-up is actually a function of component tolerances, especially those of the transistors. The collector of Q_2 decreases in value and is coupled to the base of Q_1 through C_2. This causes Q_1 to conduct less, increasing its collector voltage which is coupled back to the base of Q_2 via capacitor C_1. Eventually Q_2 is saturated and Q_1 is completely cut off, except of course, for some small insignificant leakage current. This condition only exists for a short period of time.

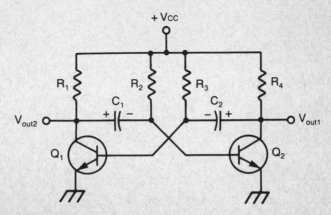

Fig. 11-22. A typical astable multivibrator circuit.

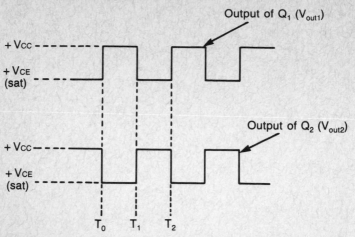

Fig. 11-23. Output waveforms of the circuit of Fig. 11-22.

Now, with Q_1 cut off, its collector voltage increases. C_1 then charges up from ground, through the base-emitter junction of Q_2, and through R_1 back to $+$ VCC. C_2 discharges through R_3 and also through Q_2 since Q_2 is now conducting. The length of discharge of C_2 is a function of the time constant of R_3C_2. Once C_2 discharges to zero volts, it begins to charge up again until at about 0.7 volts the emitter-base junction of Q_1 is forward biased causing it to begin conducting. C_1 now begins to discharge, because the collector voltage of Q_1 decreases, cutting off Q_2 for a length of time determined by R_2C_1. Once C_1 discharges to zero volts, it then begins to charge up in the opposite direction as did C_2 previously. Once this voltage reaches 0.7 volts, Q_2 starts to conduct. In the meantime C_2 has charged up to $+$ VCC and the cycle begins again.

Since R_2C_1 and R_3C_2 are the time constant components that cause charge and discharge times, thus on and off times, the frequency of oscillations can be determined from the following equation:

$$f = \frac{1.44}{3\ RC}$$

f = Frequency of operation
R = R_2 when $R_2 = R_3$
C = C_1 when $C_1 = C_2$

250

When capacitor values or resistor values are changed, frequency also changes producing various pulse widths.

An astable multivibrator using an op amp as the active device is shown in Fig. 11-24. When the circuit is initialized the op amp output will saturate to either a positive or negative value. Feedback is then applied back to both inputs via R_1 and R_2. Capacitor C_1 supplies the input signal to the inverting input while R_2 and R_3 form a voltage divider allowing a reference voltage to be applied to the noninverting input of the op amp. Capacitor C_1 charges as feedback current begins to flow. Once the voltage across C_1 is greater than the reference voltage, the inverting input causes the output polarity to reverse, charging C_1 in the opposite direction and the process begins again.

It is the value of R_1C_1 and the reference voltage that determines the output frequency of the op amp multivibrator. R_1 can be made variable to vary frequency. In any event, the output frequency is found from the following equation:

$$f = \frac{1}{2.2R_1C_1}$$

The Monostable Multivibrator

Since the monostable multivibrator has a single stable state and produces one output pulse for one input pulse, it is sometimes referred to as a one-shot multivibrator and even as a pulse stretcher. Figure 11-25 is an example of a schematic diagram of a monostable multivibrator. Other ways of connecting components can be used,

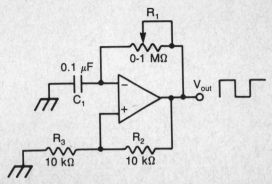

Fig. 11-24. An astable multivibrator using an op amp as the active device.

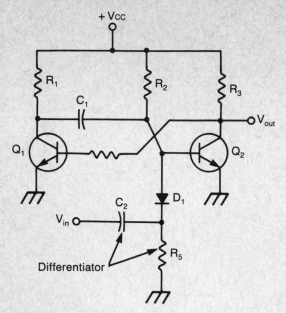

Fig. 11-25. A monostable multivibrator with a differential input.

especially without the use of the diode, but this allows you to see why this circuit is sometimes called a pulse stretcher.

In the stable state, R_2, D_1, and R_5 form a voltage divider network to forward bias and saturate Q_2 when the circuit is initially turned on. This causes collector voltage to be too low to allow Q_1 to conduct. Since Q_1 is cut off, its collector voltage, now at Vcc, charges capacitor C_1 through R_1 and the emitter-base junction of Q_2 which is turned on. The circuit remains in this state until it receives an input pulse. The input pulse and other pulses are shown in Fig. 11-26.

C_2 and R_5 form a differentiating circuit. From previous studies in this chapter you've learned that a square wave is made into a positive and negative pulse when applied to a differentiating network. This is shown again in Fig. 11-26. Only the negative portion of the pulses are applied to the base of Q_2 because only a negative going pulse forward biases D_1. This causes Q_2 to cut off. The collector voltage of Q_2 then increases causing Q_1 to begin conducting. Now the collector voltage of Q_1 decreases.

This causes C_1 to discharge through R_2, keeping Q_2 cut off for a length of time determined by the RC time constant of R_2C_1. Once C_1 discharges to zero volts, it begins to charge up again

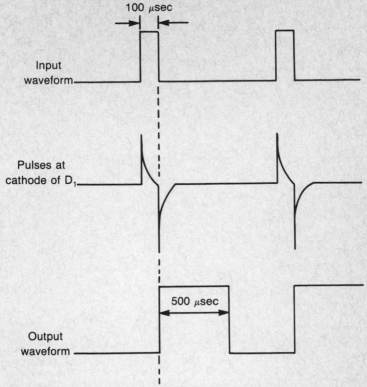

Fig. 11-26. Input and output waveforms of Fig. 11-25.

until, when it charges to 0.7 volts, it causes Q_2 to conduct. Now the collector voltage of Q_2 decreases, causing Q_1 to cut off returning the multivibrator to its stable state. This process repeats again once another pulse is received. As you can see, the output pulse has a long duration as compared to the input pulse; thus, the term pulse stretcher.

The Bistable Multivibrator

The bistable multivibrator has two stable states, two inputs, and two outputs. An input pulse to one of its inputs sets this multivibrator to one state. An input pulse to the other input resets the circuit to its other state. The circuit is sometimes referred to as a set-reset flip-flop because the set pulse flips multivibrator to one stable state while a reset pulse flops the circuit back to the original stable state. Today most multivibrators, especially bistable multivibrators called set-reset flip-flops, are found in integrated

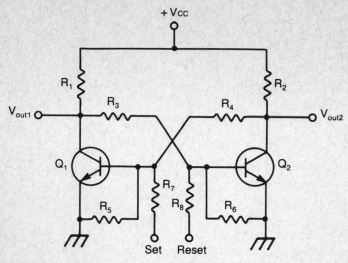

Fig. 11-27. A typical bistable multivibrator can be constructed as shown in this schematic.

circuit form. Since ICs for the most part are reserved for other texts, with some exceptions, the circuit of Fig. 11-27 using discrete components serves to show the schematic for a bistable multivibrator. Figure 11-28 shows the input signals and the output at the collector of Q_2.

When an input pulse is received at time T_0 on the set input, the output goes high and remains so until time T_1 when the reset input pulse causes the output to return to its original state. When

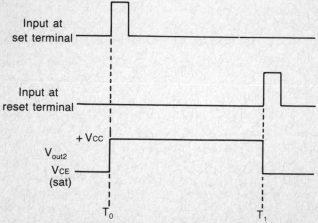

Fig. 11-28. Input and output waveforms of the bistable multivibrator of Fig. 11-27.

set, Q_1 is saturated and Q_2 is cut off and when reset, Q_1 is cut off and Q_2 is saturated giving the results shown in Fig. 11-28.

The Schmitt Trigger

The purpose of the Schmitt trigger shown in Fig. 11-29 is to convert a sine wave into a square wave. In this circuit, when the sine wave input to the base of Q_1 is below a specific positive value, Q_1 remains cut off. This causes a high positive value to be felt on the base of Q_2 causing it to saturate. The current through Q_2 come up from circuit ground and through R_3, causing R_3 to develop a large positive voltage to keep the emitter of Q_1 reverse biased, holding it in cutoff.

Once the input signal voltage rises 0.7 volts above the voltage drop across R_3, Q_1 conducts or triggers at time T_0 as shown in Fig. 11-30. Now, the collector voltage of Q_1 decreases causing the base voltage of Q_2 to decrease. Q_2 now conducts less causing the current through R_3 and therefore, the voltage drop across R_3, to decrease. This, in effect, is positive feedback forcing Q_2 to cut off and Q_1 to saturate. The output voltage is now at maximum or it can be said that it has been triggered on by turning on Q_1 at time T_0.

At time T_1, when the input signal voltage falls below a certain voltage, Q_1 is no longer saturated and its collector voltage begins to increase. (The discrepancy between Q_1 on and Q_2 off comes

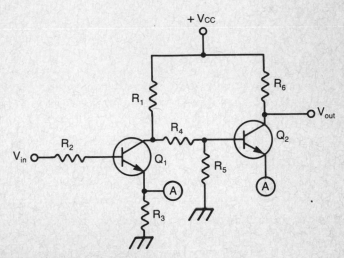

Fig. 11-29. A Schmitt trigger circuit.

255

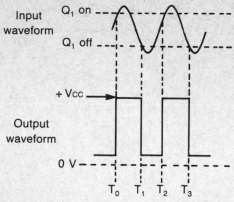

Fig. 11-30. Input and output waveforms of the circuit of Fig. 11-29.

from the fact that in reality the circuit will reset at a lower voltage than when it was turned on. This is due in part to component tolerances).Once the collector voltage of Q_1 increases, Q_2 is turned on, forcing Q_2 to quickly saturate causing the output voltage to fall just as fast. In essence, when a sine wave is applied to the input of the base of Q_1, a square wave is produced at the output of Q_2.

THE 555 TIMER

Entire chapters have been dedicated to operational circuits using the 555 timer as the active device. It is basically an integrated circuit containing circuitry enabling it to be used in a number of different ways. It can be used as an astable or monostable multivibrator and can function as a timing delay circuit and as a frequency divider. The popularity of this IC has caused a number of manufacturers to produce this device. Figure 11-31 is an example of the inside layout of the 8-pin IC 555 timer. There are also ICs that use two or even four timers in one package. As you can see from the layout, you have already studied each of these internal circuits. The basic 555 timer consists of two op amp comparators, one set-reset flip-flop, a power amplifier (PA), and two transistors, one an npn and the other a pnp. Figure 11-32 is a photo of some 555 timer ICs.

Each one of the pins on the 8-pin version has a specific function. Pins 1 and 8 are obvious. Pin 2 is used to start the timing period of the output which is activated by a low going trigger pulse. Pin 3 is the output of the power amplifier that is of a high current type enabling it to drive TTL logic circuitry. Pin 4, when reset to a low

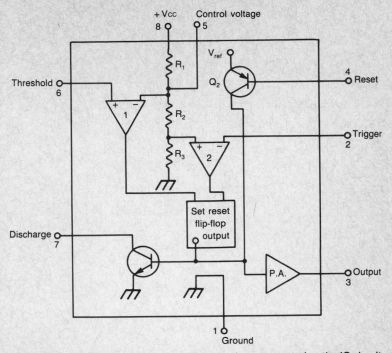

Fig. 11-31. In a simplified form, this is what may be represented as the IC circuit of a typical 555 timer.

level, disables the timer. With no input on pin 5, the reference voltage to which the threshold and trigger levels must be compared is set to 2/3VCC on the - input to comparator 1. Pin 6 forces the output of comparator 1 positive, resetting the flip-flop when the voltage on it exceeds the reference voltage. Pin 7 is connected to an external capacitor which will usually determine the time constant of the 555 timer.

In a more practical sense, Fig. 11-33 is an example of how the

Fig. 11-32. A photo of some 555 IC timers manufactured by different companies, but all carrying the number 555 (John Sedor Photography).

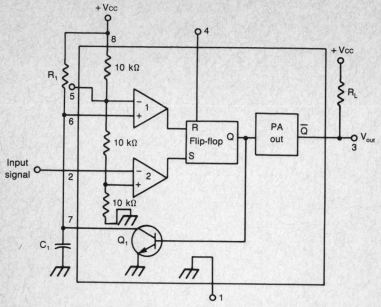

Fig. 11-33. A 555 IC configured as a monostable multivibrator.

555 timer is used as a monostable multivibrator. R_1 and C_1 control the length of the output pulse. The operation of this timer configuration can be followed by viewing Fig. 11-34. At time T_0, between negative going input pulses, the input voltage level is higher than the trigger voltage of comparator 2. The flip-flop resets causing its output Q to be at VCC. The output at pin 3 is zero volts because the PA inverts the output of the flip-flop. This causes

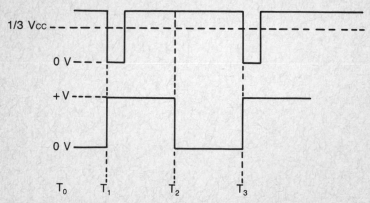

Fig. 11-34. Input and output waveforms of the 555 monostable multivibrator.

258

current flow through R_2. At this time Q_1 conducts acting as a short across C_1.

At time T_1 a negative going input pulse arrives. The input to comparator 2 now drops below the 1/3Vcc reference voltage causing the comparator to set the flip-flop. The output of the flip-flop goes to zero volts cutting off transistor Q_1, allowing C_1 to charge through R_1 to Vcc. The output of the PA is now a high level since it inverts the output of the flip-flop and remains this way until the flip-flop is reset. When the charge on the capacitor reaches $^2/_3$ the voltage of Vcc, comparator 1 switches state and resets the flip-flop. This occurs at time T_2. This causes Q_1 to conduct again, shorting C_1 and allowing it to discharge. Since at this time the flip-flop is reset, its output is inverted by the PA and as a result is seen as zero volts on pin 3.

It is the leading edge, not the trailing edge, of the input pulse that determines the start of the positive output pulse and the width of the pulse is determined by how long it takes C_1 to charge to 2/3Vcc. This time, in effect, is determined by the RC time constant of R_1C_1. In this circuit, five time constants are not necessary to charge C_1 to 2/3Vcc, just 1.1 time constants. The following equation is used to determine the pulse width of the output waveform:

$$PW = 1.1R_1C_1$$

Even if the time between input pulses is shorter than the pulse width of the output pulse, C_1 must charge to 2/3Vcc before an input pulse will cause the output pulse to drop to zero again. Therefore an output pulse could be 2 or more times greater than the input pulse resulting in frequency division. This is just one way in which the 555 timer can be used.

RAMP GENERATORS

The purpose of a ramp generator is to produce either a sawtooth, trapazoidal, or triangular waveform. The ramp generator can be an oscillator which requires no input signal voltage, or it can be a circuit that uses a transistor or an op amp as its active device that does require some sort of input signal voltage to produce a particular output. Figure 11-35 shows three of the types of ramps that are used in some electronic circuits.

A ramp, such as the sawtooth waveform, can be formed by charging a capacitor at a linear rate. This can be done by using

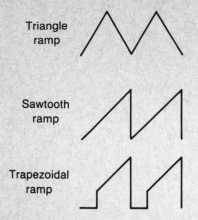

Triangle ramp

Sawtooth ramp

Trapezoidal ramp

Fig. 11-35. Basic ramps used in many of today's electronic circuits.

a constant current. A constant current flowing into or out of a capacitor causes a linear increase in the voltage across the capacitor being charged, resulting in a ramp.

The Triangle Wave Generator

The most popular method used in producing a triangle wave is by the use of integration. This is shown in Fig. 11-36, using an op amp, and in the input and output waveforms of Fig. 11-37.

Initially, at time T_0, the input waveform goes to a positive value. This causes current to flow from the output of the op amp, back through C_1 and R_1, and to the input. C_1 then charges to the polarity shown. This constant current forms a linear, negative going ramp, until at time T_1 when the input goes negative. This makes the output to go positive at a linear rate forming an inverted triangle between times T_0 and T_2.

Sawtooth Generator

The sawtooth generator is used extensively in CRT displays

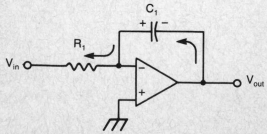

Fig. 11-36. A triangle wave generator using an op amp as the active device.

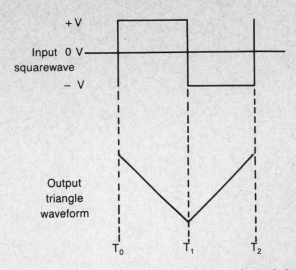

Fig. 11-37. The input and resultant output waveforms of a basic integrator.

to sweep the electron beam from one side of the screen to the other. It can also be used in an oscillator circuit to sweep it through a specific range of oscillating frequencies. Figure 11-38 is an example of a sawtooth generator using discrete components. Figure 11-39 is a timing diagram of this circuit and is a visual representation of what occurs at the output of the generator with a square wave input. If Q_1 and its surrounding components look familiar, it is because it is acting as a constant current source. It

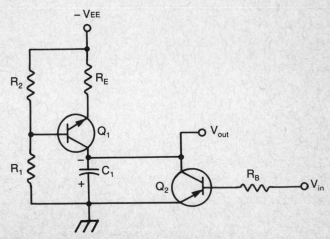

Fig. 11-38. A sawtooth generator constructed of discrete components.

261

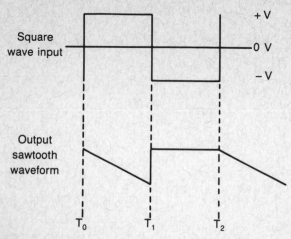

Fig. 11-39. Input and output waveforms of the sawtooth generator of Fig. 11-38.

is C_1 that forms the sawtooth waveform by charging and discharging as Q_2 turns on and off.

Looking at Fig. 11-39, at time T_0 when the input signal voltage is at its maximum amplitude, Q_2 is cut off. Current therefore flows from $-\text{VEE}$, up through Q_1 and into C_1, charging it with the polarities as shown. The constant current source charges up C_1 at a linear rate. This is shown as the negative going output ramp in Fig. 11-39. At time T_1 the input pulse swings negative, turning on Q_2 which acts like a short across C_1. C_1 then discharges rapidly through Q_2 as shown in the output waveform. This negative portion of the input square wave keeps Q_2 turned on and saturated and the current from Q_1 flowing around C_1 and through Q_2 keeping C_1 from charging up again. This process of ramping is begun again at time T_2 when the input signal voltage swings positive.

As you can see, the frequency of the output sawtooth waveform is a function of the frequency of the input square wave and the slope of the ramp is a function of the size of C_1 and the value of current through that capacitor.

Chapter 12

Power Supplies

In most types of electronic circuits, a dc supply voltage is necessary for proper operation. Usually this is accomplished by converting power from an incoming ac line to direct current using a circuit called a power supply. There are times when the needed direct current can be supplied by a battery, but in most cases converting ac to dc is required. This chapter deals with some of the ways in which conversion takes place. In some cases operational amplifiers are discussed since they are found in many circuits that regulate the dc voltage once it has been converted from an ac voltage to an unregulated dc voltage.

RECTIFIER CIRCUITS

A rectifier circuit is the first step required in changing an ac voltage into a dc voltage. It doesn't actually convert ac into dc, but rather into a pulsating dc. This is, at least, a step in the right direction.

There are several types of rectifier circuits, the half-wave rectifier, full-wave rectifier, and bridge rectifier. There are also rectifier circuits used for single phase operation and for polyphase operation. In this chapter, only single phase rectifier circuits are discussed.

Half-Wave Rectifier

The half-wave rectifier is the simplest of the rectifier circuits.

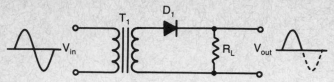

Fig. 12-1. A typical half-wave rectifier circuit.

Figure 12-1 shows a half-wave rectifier. A single diode is connected in series with the secondary of transformer T_1. A diode conducts in one direction only, once its junction voltage of about 0.7 volts is overcome, so the diode acts like a closed switch once it begins to conduct.

When the ac input voltage is on its positive voltage cycle, the diode conducts and allows this positive alternation, or pulse, to appear at the output across the load resistor R_L. On the negative alternation of the input ac voltage, the diode cuts off and no current flows through the load resistor; no output voltage is developed. The output looks like a series of positive pulses. This type of output is referred to as pulsating dc and is not useful in most applications requiring a dc voltage for operation. But again, it is a step in the right direction, since current is flowing in one direction only.

In a normal sine wave, the average value of voltage is zero volts because the value of positive voltage and the value of negative voltage nullify each other, or average out to zero volts. In a half-wave rectifier circuit only a positive voltage value is obtained because, in reality, the bottom portion of the input voltage sine wave has been clipped by action of the diode. However, an average value of the positive pulses can still be found, and does exist.

Ac voltage values that have been rectified are still expressed in terms of their effective (rms), peak, and average values. A standard 115 Vac sine wave commonly found in a wall outlet is shown in Fig. 12-2. In this figure, the rms value (that value that an ac

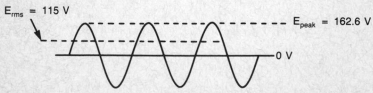

Fig. 12-2. A typical 115-Vac sine wave showing RMS and peak value comparisons.

264

voltmeter measures) is shown to be 115 Vac. The peak value is found from the equation:

$$E_{peak} = E_{rms} \times 1.414$$

Therefore, in this example:

$$E_{peak} = 115 \times 1.414 = 162.6 \text{ Vac}$$

In Fig. 12-3, the negative portion of the sine wave has been clipped by a half-wave rectifier circuit. The average voltage of the sine wave is no longer zero volts, since now only a positive value exists. Therefore, the average voltage, E_{ave}, is found from the following equation:

$$E_{ave} = E_{peak} \times 0.318$$

Therefore, in the example of Fig. 12-3:

$$E_{ave} = E_{peak} \times 0.318$$
$$E_{ave} = 162.6 \times 0.318$$
$$E_{ave} = +51.7 \text{ Vdc}$$

In this case, the small voltage drop across the diode of about 0.7 volts is very insignificant and is therefore disregarded. Also, since the frequency of the output pulses is the same as the input pulses, a 60 Hz ac input sine wave produces pulses with a frequency of 60 Hz, called the ripple frequency of the output dc waveform.

Full-Wave Rectifier

Since a half-wave rectifier only rectifies half of the incoming ac voltage, there are some serious disadvantages incurred with that

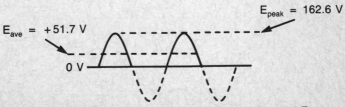

Fig. 12-3. The output of a half-wave rectifier showing E_{ave} and E_{peak}.

type of arrangement. The full-wave rectifier overcomes filtering problems associated with the half-wave rectifier and is shown in Fig. 12-4.

Notice that there are now two diodes in the output or transformer secondary circuit and that the secondary windings of the transformer are center-tapped. A grounded center-tap produces voltages at the opposite ends of the secondary windings (top and bottom halves) that are 180° out of phase with each other.

Examining the circuit of Fig. 12-4, the following action occurs during the positive alternation of the ac input waveform. The anode of D_1 is positive while the anode of D_2 is negative. This means that D_1 conducts current from the center-tap of the transformer, through the load, through D_1, and to the top of the transformer secondary. A positive half cycle is developed across the load. During the negative half cycle of the ac input voltage waveform, the anode of D_2 is positive while the anode of D_1 is negative. D_2 conducts from the center tap, through the load, through D_2 and back to the bottom of the transformer secondary. Another positive half cycle is developed across the load. In effect, a positive pulse is developed during both positive and negative alternations of the input ac voltage waveform.

To determine the rms value at the output of the secondary, assume a transformer with a 1:1 turns ratio. The output across the entire secondary is 115 Vac for an ac input of 115 Vac. Since the transformer is center-tapped, the voltage across each half, from center tap to top or from center tap to bottom, of the secondary is one half of 115 Vac, or 57.5 Vac, rms. The peak value can be found as follows:

$$E_{peak} = E_{rms} \times 1.414$$
$$E_{peak} = 57.5 \times 1.414$$
$$E_{peak} = 81.3 \text{ Vac}$$

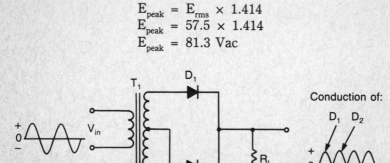

Fig. 12-4. A full-wave bridge rectifier showing input and output waveforms.

266

Now that the peak voltage, E_{peak}, is known for each pulse, the average voltage, E_{ave}, can be found in the secondary by the same equation as for the half-wave rectifier, but now must be multiplied by 2 since pulses occur twice as often in this full-wave rectifier. Therefore, the equation states:

$$E_{ave} = 2 \times E_{peak} \times 0.318$$
$$E_{ave} = E_{peak} \times 0.636$$
$$E_{ave} = 81.3 \times 0.636$$
$$E_{ave} = 51.7 \text{ Vdc}$$

This average voltage is still the same in the output of the full-wave rectifier as it is in the output of the half-wave rectifier with equal input voltages. The advantage is that now the ripple frequency is 120 Hz, not 60 Hz, making it easier to turn this pulsating dc into a smooth dc needed for electronic circuit applications. Also, in the full-wave rectifier circuit, the average voltage is closer in value to the peak voltage than in a half-wave rectifier. In fact, the peak voltage of the full-wave rectifier is only half the value of a half-wave rectifier. This is one disadvantage of the full-wave rectifier.

Bridge Rectifier

The biggest advantage of the bridge rectifier is that it eliminates the fact that in either the full-wave or half-wave rectifier circuits, the average voltage remains the same. It combines the best features of both previous rectifier circuits by having the amplitude of the output pulse the same as that of the half-wave rectifier while maintaining a ripple frequency of 120 Hz like that in the full-wave rectifier. An example of a bridge rectifier circuit and its current flow, and thus output voltage development on both positive and negative cycles of the ac input voltage, is shown in Fig. 12-5.

In this schematic, an input transformer is being used. However, a transformer is really only needed for isolation purposes or when step up or step down in ac input voltage is required. The ac input voltage can actually be tied directly to the top and bottom portions of the bridge itself, adding another advantage to this type of rectification.

During the positive half cycle of the ac input sine wave, the current flow is from the bottom of the transformer secondary through D_1, up through the load, through D_2, and back to the top of the transformer secondary. At this time D_3 and D_4 are reverse biased and do not conduct. This action develops a positive pulse

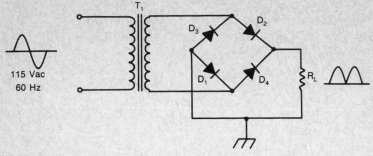

Fig. 12-5. A bridge rectifier showing input and output waveforms.

across the output load. During the negative half cycle of the ac input sine wave, the current flow is from the top of the secondary of the transformer through D_3, up through the load (notice this current flow is in the same direction through the load as before), through D_4, and back to the bottom of the secondary of the transformer.

During both the positive and negative half cycles of the ac input voltage, the entire secondary voltage is developed across the output load. Figure 12-6 is an example of the average output voltage developed across the load using a bridge rectifier circuit. Earlier you saw that an rms voltage of 115 volts actually produces a peak voltage of 162.6 volts. The average voltage of the bridge rectifier circuit is found using the same equation used for finding the average voltage in a full-wave rectifier circuit. For an average voltage of 115 volts:

$$E_{ave} = E_{peak} \times 0.636$$
$$E_{ave} = 162.6 \text{ V} \times 0.636$$
$$E_{ave} = 103.4 \text{ Vdc}$$

Notice that this is twice the average voltage of the full-wave rectifier, but with the same ripple frequency.

FILTER CIRCUITS

The pulsating dc voltage that is a product of all three of the

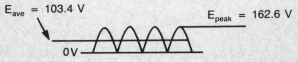

Fig. 12-6. The average voltage of the output of a bridge rectifier as compared to the peak voltage.

rectifiers is simply not suitable for use in most electronic circuits requiring a dc voltage. This dc voltage needs to be as steady state or smooth as possible, just like the dc from a battery. One way of smoothing out a pulsating dc voltage is by using a large capacitor across the output of the rectifier circuit. This is called a filter capacitor, because its purpose is to filter out the pulses, or ripples, that are produced by these circuits.

Capacitive Input

Figure 12-7 is an example of a half-wave rectifier with a capacitor across the output load. Notice that the output wave form begins to flatten out with the use of a large capacitor. The action that takes place in this half-wave circuit is one that allows the capacitor to charge up to the peak voltage when diode D_1 conducts and current flows through the load. Once the ac input cycle causes the diode to cut off, C_1 tries to discharge, but its only discharge path is through the load. If C_1 and the load are large enough, the capacitor does not have a chance to completely discharge through the load before D_1 conducts again, charging up C_1. Therefore the output voltage is actually the charge on C_1 from one positive input half-cycle to the next. The discharge and charge times of C_1 are a function of the RC time constant of $R_L C_1$. This capacitor also raises the output average voltage because now the output voltage pulses never reach zero volts, but only decrease to a point in the capacitor's discharge time where the capacitor begins to charge once again. Since the capacitor is actually supplying current through the load when D_1 is not conducting, a larger capacitance value is necessary when the current through the load is very high.

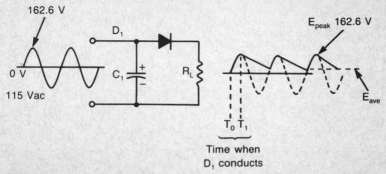

Fig. 12-7. The output waveform is a result of using a capacitor on the output of this half-wave rectifier.

Figure 12-8 is an example of the output waveform of a full-wave rectifier circuit using a filter capacitor across the output load. This is also the same waveform that is present at the output of a bridge rectifier circuit. The advantage here is that the capacitor doesn't have much time to discharge before the next pulse arrives. In fact, with twice the pulses, a large capacitor can raise the average voltage very close to E_{peak} resulting in a very smooth dc output voltage.

To describe just how good a job a capacitor is doing in smoothing out a pulsating dc voltage, the term percent ripple is used.

$$\text{Percent ripple} = \frac{\text{rms of ripple}}{E_{ave}} \times 100$$

In this equation the rms of ripple is found by looking at the very top portion of the output waveform and measuring the peak-to-peak voltage of the ripple voltage. As an example, Fig. 12-8 shows an expanded top portion of the output of a filtered full-wave or bridge rectifier circuit. The peak-to-peak value of the ripple voltage, as can be seen, is 2 volts. Its peak voltage is 1 volt. The rms value is 0.707 of the peak voltage, or in this case, 0.707 volts. Using the percent ripple equation yields:

$$\text{Percent ripple} = \frac{\text{rms of ripple}}{E_{ave}} \times 100$$

$$\text{Percent ripple} = \frac{0.707 \text{ V}}{11 \text{ V}} \times 100$$

$$\text{Percent ripple} = 0.064 \times 100$$
$$\text{Percent ripple} = 6.4\%$$

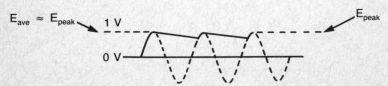

Fig. 12-8. The output waveform of a large capacitor on the output of a full-wave rectifier, where Eave approaches Epeak.

In most cases, the load resistance is really a function of circuit design and therefore not readily variable. The filter capacitor, however, can easily be changed and can be calculated for a full-wave or bridge rectifier circuit by using the following equation:

$$C_{min} = \frac{1}{2.828 \times K \times R_L \times f}$$

Cmin = Minimum capacitance required in farads
K = Percent ripple expressed as a decimal
R_L = Resistance of load in ohms
f = Frequency of ripple in hertz

RC Filters

Capacitor values in the simple capacitor circuits just discussed can reach very high values and therefore become somewhat prohibitive from a financial and mechanical standpoint. Other methods are used that can be space saving and more cost effective in the long run. One of these methods is the use of an RC filter as shown in Fig. 12-9.

This is a full-wave rectifier circuit with an RC filter instead of the large single capacitor. C_1 is basically a filter as was used previously. However, following C_1 is an RC network consisting of $R_1 C_2$. R_1 is a small value preventing voltage loss to the load while C_2 serves as an infinite impedance to dc allowing the dc to be passed to the load. While offering this high impedance to dc, it offers a very low impedance to ac. The ac ripple that is seen at the output of C_1 is now passing on to a voltage divider network made up of R_1 and C_2 to ground. Since R_1 is made considerably larger than the reactance of C_2 at the ripple frequency, most of the ripple frequency is dropped across R_1 with almost no ripple seen across C_2 and the output load. The voltage across the output looks quite smooth and steady state.

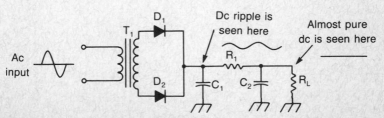

Fig. 12-9. A full-wave rectifier circuit with an RC filter (pi, π, filter) on the output to reduce power supply ripple.

271

LC Filters

Although the LC filter is probably bulkier and more expensive than the RC filter, it does offer some distinct advantages. Figure 12-10 is an LC filter circuit on the output of a full-wave rectifier.

It is very much like the RC filter except that the resistor has been replaced with an inductor wound on an iron core. This inductor offers a reactance to ac at the ripple frequency, but does not block dc. L_1 and C_2 form a voltage divider network with C_2 offering a very low reactance to ac ripple. Again, the reactance of L_1 is much higher than the reactance of C_2, L_1 drops almost all of the ripple voltage while only a small ripple voltage is felt across C_2 and consequently, the output load.

Since L_1 offers almost no resistance to dc, no dc voltage is lost across it, resulting in more voltage to the load, and also very little heat dissipation from the inductor, unlike the resistor. Another advantage is that current through the inductor tends to resist any changes in the magnitude of current flowing through it.

VOLTAGE MULTIPLIERS

In some instances it may be necessary to increase the output voltage of a rectifier circuit by several times. Usually a step up transformer is used to increase the output voltage from a rectifier circuit, but there are easier, more cost-effective methods in stepping up the output dc voltage of rectifier circuits. These voltages can be doubled or even tripled.

Half-Wave Voltage Doubler

A half-wave voltage doubler is shown in Fig. 12-11. Assume an input ac sine wave of 115 Vac and 60 Hz. This means a peak voltage of approximately 162 volts. During the negative half-cycle of the input ac waveform, D_1 conducts charging C_1 to the peak

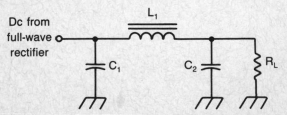

Fig. 12-10. Another pi filter at the output of a full-wave bridge rectifier, this time using a coil (choke) and capacitors.

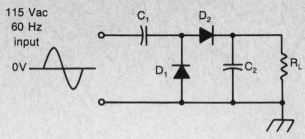

Fig. 12-11. A simple half-wave voltage doubler that produces twice the input voltage.

voltage of the ac input sine wave. There is no path for discharge so C_1 remains charged until the next half-cycle of the input waveform. This means that the point between C_1 and D_1 is now charged to 162 volts in reference to ground. During the positive half-cycle of the input ac waveform, D_2 conducts allowing C_2 to charge up. The voltage across C_2 is now the peak voltage of the negative half-cycle plus the charge on C_1, for a total of 324 volts. You can see that C_2 must have a voltage rating that is equivalent to the output voltage of the voltage doubler, a distinct disadvantage.

Full-Wave Voltage Doubler

The full-wave voltage doubler shown in Fig. 12-12 eliminates the filtering difficulties found with the half-wave doubler and the capacitor voltage rating problem. On the positive half cycle, D_1 conducts, allowing C_1 to charge to the peak voltage of the input

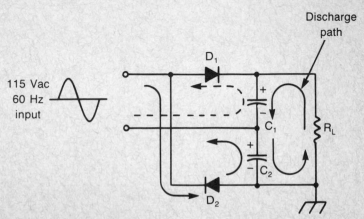

Fig. 12-12. A full-wave voltage doubler with charge and discharge paths shown.

273

ac voltage. During the negative half cycle, D_2 conducts and now C_2 charges to the peak voltage of the input ac voltage. D_1 and D_2 now only conduct during the peaks of the ac input waveform because the charge on the capacitors at times other than during peaks, keeps the diodes reverse biased. The capacitors, charged series aiding, discharge across the load at twice the voltage of the ac input voltage. Since neither capacitor is charged to twice the input ac voltage, the rating voltage is not a problem.

The Voltage Tripler

Voltages can be not only doubled, but can also be tripled by the use of a circuit called a voltage tripler, shown in Fig. 12-13. This circuit operates for a period of one and a half cycles before tripling of the ac input voltage takes place.

During the first positive half cycle, C_1 charges up to the ac peak voltage through D_1 which is forward biased at that time. During the peak of the negative half cycle, which is the next half cycle, the top of C_2 is negative with respect to ground. It is a negative voltage equal in value to the peak voltage of the ac input voltage. This negative voltage forward biases D_2 and current flows from C_2 to the top of C_1. C_2 is charged to twice the peak voltage of the ac input voltage with respect to ground. During the positive peak of the next half cycle, C_3 charges up through D_3. This is an additional diode-capacitor section that adds the value of the peak ac input voltage to the already doubled value of E_{peak}. The output across the load is now three times the value of the input peak ac voltage. Additional diode-capacitor sections can be added to construct quadruplers. But, as multiplication continues, regulation

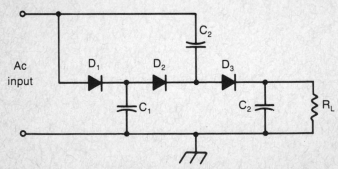

Fig. 12-13. To better understand this voltage tripler circuit, indicate charge and discharge paths on this drawing while reading the circuit description in the text.

decreases as does the current supplied by this type of multiplication circuitry.

VOLTAGE REGULATION

Thus far, the rectifier circuits discussed offer very little in the way of providing a constant output voltage when various currents are drawn through the output load. Even the ac linear input voltage is not a steady 115 Vac and the converted dc output voltage of a rectifier circuit is also unsteady or unregulated adding to the already changing value in load resistance. Probably the greatest argument for dc supply voltage regulation is the fact that most electronic circuits are designed to operate within a very narrow range of dc voltages. In some circuits a variation of more than 0.1 percent in the dc supply voltage causes improper operation of the circuit. Voltage regulation is therefore an important aspect in power supply considerations.

The percentage of regulation is a measure of how effective a power supply is in maintaining a particular output voltage for which it was designed. It can be found from the following equation:

$$\text{Percent of reg.} = \frac{E_{nl} - E_{fl}}{E_{fl}} \times 100$$

E_{nl} = output voltage under no load conditions
E_{fl} = output voltage under full load conditions

As an example, if the output of a dc power supply is 20 volts when no current is flowing through the load, and that same output drops to 15 volts when the maximum current flows through the load, then the percentage of regulation is:

$$\text{Percent of reg.} = \frac{20 - 15}{15} \times 100$$

$$\text{Percent of reg.} = \frac{5}{15} \times 100$$

$$\text{Percent of reg.} = 0.333 \times 100 = 33.3\%$$

Zener Voltage Regulator

One of the simplest voltage regulators is the zener voltage regulator, which uses the zener diode for voltage regulation. The characteristic that makes this diode useful for this function is the fact that once reverse voltage breakdown occurs, the voltage across the diode remains relatively constant over a wide range of currents.

An example of a basic zener diode regulator is shown in Fig. 12-14. This circuit is added between the output of a rectifier's filter circuit and the output load. Its input is an unregulated dc voltage. The zener diode is reverse biased. The zener is chosen to have a breakdown voltage equal to the required output voltage of the power supply. This circuit is sometimes referred to as a shunt regulator because the zener diode shunts the output load.

The total current, I_T, is made up of the current flowing through the zener diode, I_Z, and the current flowing through the load, I_L. R_S is chosen so that I_Z stays at a sufficient level to keep the zener in its breakdown region. It must also allow the current to flow through the load that is required.

If the load resistance increases, I_L decreases. VRL tries to increase but the zener diode conducts harder so that the total current flowing through R_S remains essentially constant. The same voltage is maintained across R_S and the load. When I_L increases, I_Z tends to decrease. As long as I_Z stays at a level sufficient to keep the zener in its breakdown region, the output voltage remains constant because the load is in parallel with the zener diode.

As you can see, R_S is an important part of the zener regulator circuitry. To determine the value of R_S, the following equation is used.

$$R_S = \frac{E_{in(min)} - E_Z}{(1.1) \, I_{Lmax}}$$

R_S = The zener series limiting resistor
$E_{in(min)}$ = Minimum input voltage
E_Z = Zener diode voltage
I_{Lmax} = Maximum load current

Also keep in mind that zener diodes have wattage ratings just like resistors. A wattage rating, or power dissipation, must be determined to prevent damaging the zener diode. The maximum

276

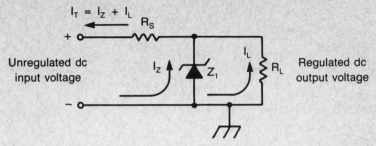

Fig. 12-14. A simple basic zener diode voltage regulator.

power dissipation can be determined from the following equation:

$$P_{Zmax} = E_Z \left(\frac{E_{in(max)} - E_Z}{R_S} - I_{Lmin} \right)$$

P_{Zmax} = Maximum zener power dissipation
E_Z = Zener diode voltage
$E_{in(max)}$ = Maximum input voltage
I_{Lmin} = Minimum load current

Series Voltage Regulator

One type of series voltage regulator is shown in Fig. 12-15. The transistor, Q_1, is in series between the input and output voltage.

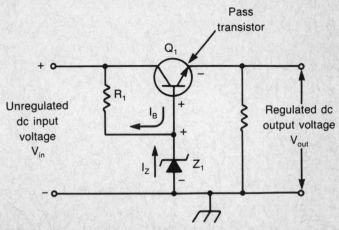

Fig. 12-15. A basic series voltage regulator using an NPN transistor, in an emitter follower configuration, as a pass transistor.

The zener diode is in the base circuit of Q_1. Q_1 acts as an emitter follower, so this circuit is sometimes referred to as an emitter follower regulator. The input is an unregulated dc voltage while the load is connected between emitter and ground. The output voltage is therefore equal to the zener voltage minus the small voltage drop across the emitter-base junction of Q_1.

If the input voltage tries to increase, the output voltage tries to increase also. Since the base of Q_1 is set at a specific voltage an increase in voltage at the emitter of Q_1 causes Q_1 to conduct less (it tends to reverse bias the emitter-base junction) acting as an increasing resistance between V_{in} and V_{out}. In effect, Q_1 acts as an automatic variable resistor between V_{in} and V_{out}. This means only a very little increase in V_{out} because most of the increase in V_{in} is dropped across Q_1. This is also known as a series-pass regulator. In this sense Q_1 is the active control device called a pass transistor.

Feedback Voltage Regulator

A feedback voltage regulator is one in which part of the output voltage is fed back to a device to control the output voltage itself. Figure 12-16 is a feedback voltage regulator circuit. R_1, R_2, and R_3 form a sensing or sampling network to detect changes in the output voltage. Q_2 acts as the actual error detector and amplifier.

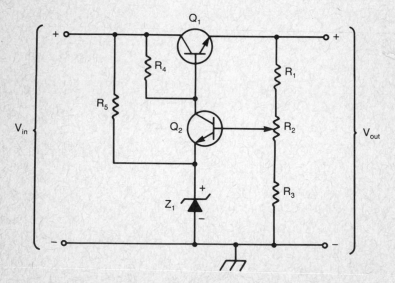

Fig. 12-16. A typical feedback voltage regulator circuit.

Any change in output voltage is felt at the base of Q_2 and further amplified, controlling the conduction of Q_1. R_5 and D_1 provide the reference voltage for the emitter of Q_2 while R_4 is used as a load resistor for the collector of Q_2 and the base biasing resistor for Q_1.

Op Amp Voltage Regulator

For better sensitivity and a faster response time to changes in output voltage, an op amp regulator is often used. The transistor of Q_2 in Fig. 12-16 has been replaced by an operational amplifier in Fig. 12-17. Here, the zener diode and R_S are used to set a reference voltage level to the noninverting input of the op amp. The feedback voltage is sensed at the inverting input of the op amp. The op amp now acts like a comparator with its output controlling the operation of pass transistor Q_1.

CURRENT REGULATION

The purpose of a current regulator is to maintain a constant current through a given output load impedance in spite of a change in that output load impedance. Basically a current regulator is a circuit providing a constant current source. These circuits were discussed earlier and used a transistor and associated components for current regulation.

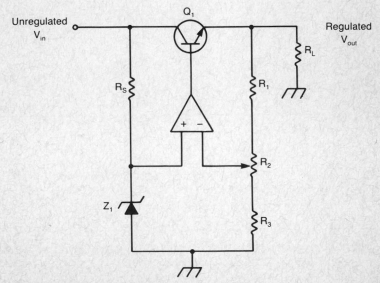

Fig. 12-17. An op amp controlled voltage regulator.

One of the more popular devices used in current regulation is the operational amplifier. As you study electronics you'll see more and more uses for this device. Figure 12-18 is an example of an op amp current regulator. I_f must be equal to I_{in} since an assumed ideal op amp has current neither entering nor leaving the inverting input. Since I_{in} is a function of V_Z divided by R_{in} and these two parameters are constant, then I_{in} must be constant and independent of any variations in R_f. R_f in this circuit is actually R_L, or the load. If there is a change in R_L and thus a change in gain of the circuit, V_{out} of the op amp changes also, maintaining a constant level of load current I_f, which is actually the load current.

POWER SUPPLY CIRCUITS

Thus far, the power supply circuits studied have provided a constant voltage output for a varying input voltage or constant current output for changes in load impedance. Filter circuits have also been covered, but there is still another important aspect to power supplies: protecting the power supply if the output load shorts and draws an excessive amount of current. The current must be limited when it exceeds a specified level or the power supply components will be damaged.

A number of protection devices are used on the output of a power supply. This may be something as simple as a fuse, which contains a thin wire that melts when a certain current level is exceeded. Another device is the circuit breaker which acts as an automatic switch that shuts off when sensing too much current

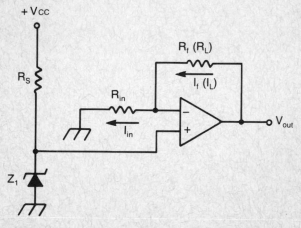

Fig. 12-18. An op amp current regulator.

through it. The advantage of the circuit breaker over the fuse is that it can be reset or turned back on once the excessive current has been limited and it comes in different current ratings like a fuse. The simple fuse however, must be replaced once it has burned open.

The problem with both of these devices is that they may not be fast enough in limiting current. In fact, in many cases a pass transistor or other power supply output device might be damaged before the devices can work to protect them. Most power supplies still incorporate fuses and circuit breakers, but increase the margin of protection by the addition of an electronic protection device that operates with more sensitivity and speed than the fuse and circuit breaker.

Overload Protection

Overloading a power supply can be caused by a short in the output of the supply, such as in the load, or by having too many devices being powered by the same power supply. At the same time the load must be protected from too much voltage if the power supply should fail to regulate its output voltage properly. Too much voltage to the load can result if a pass transistor in a series voltage regulator should short. Figure 12-19 is an example of how to protect the load from overvoltage using an op amp, zener diode, and SCR. This is called a crowbar circuit which consists of an SCR connected directly across the load. The use of an op amp increases the sensitivity of the circuit.

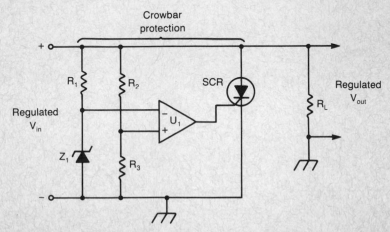

Fig. 12-19. Overload protection of a regulated dc power supply using a crowbar circuit.

If the output of the power supply exceeds the voltage of the reference voltage at the inverting input of the op amp comparator, the output of the comparator generates a positive voltage to turn on the SCR, causing it to conduct. The SCR now acts as an almost short circuit across the load and its resistance and thus voltage drops are reduced to a very low level. While the load is now protected, a short in the output like this causes the circuit breaker to kick or the fuse to blow. The SCR in this instance must be capable of handling the large current flow the instant it conducts.

Three-Terminal Regulators

A popular type of voltage regulator is the three-terminal regulator shown in Fig. 12-20. This is actually an IC voltage regulator. The chip of the IC is mounted in a TO220 transistor package that you may recognize. This allows the regulator to be mounted on a PC board or the chassis in which the other electronic components of the supply are mounted.

There are a number of three-terminal regulators available on the market today made by several different manufacturers. These regulators are also designed to produce positive or negative regulated voltages. Of course, their inputs must be of the proper

Fig. 12-20. A typical 3-terminal voltage regulator, the UC117, from Unitrode (courtesy Unitrode Corporation).

Fig. 12-21. Three terminal voltage regulators with voltage outputs of + and − 5Vdc and + and − 12Vdc.

polarity. Figure 12-21 is an example of both 5 volt and 12 volt positive and negative voltage regulators. Notice that their outside appearance is identical. The numbering on the package indicates whether the device is a 5 volt regulator or a 12 volt regulator, and whether it regulates a positive or negative voltage. In general, positive three-terminal voltage regulators contain a 78 in their numbering system while negative regulators contain a 79 in theirs. As an example, a + 5 volt three-terminal regulator has the number 7805 on it, while a − 12 volt regulator has a 7912 as its number label. Sometimes the first two digits are separated from the last two digits with one or more letters.

Figure 12-22 is a circuit using a + 5 voltage regulator. Notice the simplicity of the circuit. All that is required is a voltage input of between + 10 volts and + 35 volts. Normally the input and output sections of the regulator circuit require bypass capacitors. These capacitors shunt to ground any feedback oscillations that may be produced from lead inductance if the device is mounted more than 4 inches away from the filter circuit of the power supply.

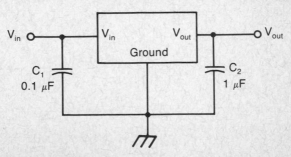

Fig. 12-22. A simplified schematic of a three terminal voltage regulator.

Programmable Regulators

None of the three-terminal voltage regulators thus far discussed can supply an output voltage other than what they are rated for except by perhaps changing the ground reference. There are, however, voltage regulators that are adjustable, or programmable, meaning that for a specific voltage in, a specific voltage output can be obtained. Figure 12-23 is an example of a Unitrode Corporation UC117, 1.5 A, three-terminal adjustable positive voltage regulator.

In this circuit, the output is adjustable from 1.2 volts to 37 volts, depending on the proper input voltage. This device also contains overload protection consisting of current limiting and thermal shutdown in the event the device overheats or ambient temperature exceeds certain limits. Thermal shutdown is very effective because once the heat diminishes, the device is once again operational. In addition, the UC117 offers exceptional line and load voltage regulation, a requisite for all voltage regulators.

SWITCHING REGULATORS

A switching regulator is a circuit that varies the duty cycle of a power pass transistor switch to provide for a regulated dc output from an unregulated dc input voltage. Unlike the typical series regulator which uses the active device as a variable resistance, this regulator maintains a high efficiency, particularly for low voltage, high current supplies and controls the output voltage by using a pass transistor as a variable duty cycle switch to produce an output voltage. This high efficiency is also maintained because it consumes considerably less power.

The Dc Switching Regulator

Figure 12-24 is an example of a dc switching voltage regulator

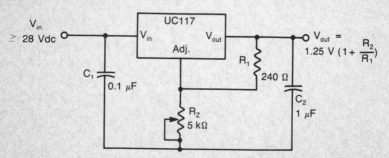

Fig. 12-23. A typical 3-terminal voltage regulator using a Unitrode Corporation UC117 voltage regulator.

284

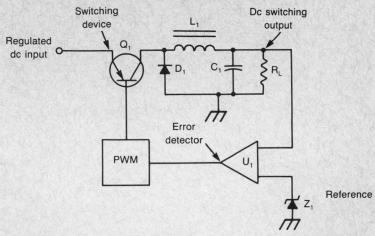

Fig. 12-24. A typical switching regulator circuit.

circuit. The op amp is used as a comparator and error detector. Any difference (error) between the reference voltage established by Z_1 and the output voltage of the power supply generates an output from the comparator that is fed to the pulse width (duty cycle) modulator. In effect, the pulse width modulator turns the switching transistor Q_1 on or off. This transistor is biased so that it is either completely on (saturated) or completely off (cut off) by the pulses on its base from the pulse width modulator.

The Pulse Width Modulator

One way of controlling switching regulators is by the use of a pulse width modulator or PWM. A popular PWM is the 1524 series manufactured by several manufacturers. Unitrode Corporation also makes the 1524 called a UC1524 shown in Fig. 12-25. As you can see this is a 16-pin IC that performs all of the functions necessary for pulse width modulation. In fact, Unitrode was the first to develop the 1524 in 1976. The PWM is also referred to as an IC control chip since it controls the switching transistor.

In the circuit of Fig. 12-24, the low pass filter of L_1C_1 converts the average dc output from the collector of Q_1 (remember, the output from Q_1 is a series of pulses) to a very smooth dc output voltage. D_1 allows the back emf of the inductor to discharge when Q_1 is cut off.

TROUBLESHOOTING POWER SUPPLIES

A basic understanding of how power supplies operate is a

Fig. 12-25. A pulse width modulator (PWM), using the UC1524, in IC form (courtesy Unitrode Corporation).

prerequisite to properly troubleshooting any power supply. There are also troubleshooting procedures that make it much easier to find the problem. These procedures are basically a process of elimination in determining what is working and what is not.

A block diagram of a power supply is shown in Fig. 12-26. When a voltmeter is not handy, a simple method of finding out which section of a shorted supply is bad is by starting at the left and working to the right. This method requires replacing the fuse each time it blows because, until the shorted section is found and isolated, the fuse will continue to blow. This method can be used to localize the shorted section of the power supply. Once that section

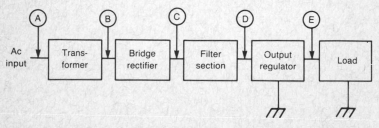

Fig. 12-26. A block diagram of a typical power supply.

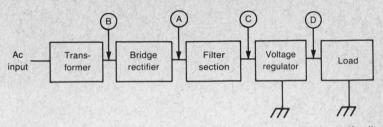

Fig. 12-27. A block diagram of a typical power supply used in the half-split method of troubleshooting.

is isolated, further troubleshooting is necessary within the stage itself.

A better method is to use a variable input transformer and an ammeter in series with the input ac voltage to the first stage of the power supply. As the ac input voltage is brought up slowly, the ammeter indicates an excessive amount of current is being drawn. This eliminates the possible burning up of components if the fuse is replaced and does not blow.

Another method of troubleshooting is the half-split method. This involves splitting the circuit in two, and narrowing the problem down to one half of the supply or the other. Figure 12-27 is a power supply block diagram using this method. Notice that the supply is first split at point A to divide it into a left half and a right half. If the supply still overloads and blows the fuse, the problem is either with the transformer or the bridge rectifier. If you break the circuit at point B and the short is still there, the problem is with the input transformer. It can't be to the right of A because these have been disconnected and therefore eliminated as possible causes of overloading. As you can see, troubleshooting in most cases can simply be a process of elimination.

Troubleshooting switching power supplies requires a knowledge of how switching supplies operate. Usually a schematic is almost essential in troubleshooting these types of supplies and is beyond the scope of this text.

Chapter 13

Integrated Circuit Technology

One of the most interesting developments in the field of solid state electronics has been that of the integrated circuit, or IC. Originally ICs were very simple linear or digital solid state integrated circuits without very much complexity, as far as circuit content is concerned. But today, ICs perform a myriad of functions and can be found in almost every piece of electronics equipment, home entertainment system, or home appliance.

There has been tremendous growth in manufacturing techniques over the last few years allowing more and more circuitry to be packaged into an area smaller than the size of a baby's thumbnail. And even these small chips can house enough circuitry to accomplish mathematical functions that years ago would have taken a room full of electronics tubes and transistors to accomplish.

There are entire texts devoted to integrated circuit technology. This chapter does not attempt to duplicate all of the material in those texts; instead, it deals with some highlights considered essential for the understanding of integrated circuit technology.

ADVANTAGES OF ICs

Probably the biggest advantage of the IC is its small size compared to other solid state devices such as the transistor or diode. In fact, hundreds of transistors, diodes, resistors, and capacitors can be found within one integrated circuit. Also, as far as manufacturing techniques are concerned, integrated circuits are made in virtually the same manner that transistors are made. This means that a substantial cost savings can be realized by using ICs rather than discrete components. In many cases a single IC that can do the same job as a dozen transistors, diodes, and capacitors costs far less than an individual transistor alone.

Using ICs results also in using far fewer parts in the assembly of some electronic circuits, again reducing cost and time in assembly of electronics equipment. Just think of the digital watch that you may be wearing. That watch contains a single IC that performs the functions of providing hours, minutes, and seconds, usually the day and month, and in some cases an alarm function to remind you of an important appointment. Some watches even play a melody and/or a game. All of these functions are contained in a single IC chip which may contain thousands of transistors, resistors, diodes, and capacitors.

DISADVANTAGES OF ICs

With all of the advantages stated above it may seem that there could not possibly be any disadvantages to this wonder of the electronic's age. The first disadvantage however, is that ICs can operate only with very small voltages and currents. They are usually used in the processing of information and for supplying signals that are used to control other devices that do handle large voltages and currents. Since they are low voltage, low current devices, they also have low power dissipation, usually less than a watt. Some MOS ICs actually operate in the microampere range.

One more disadvantage of the integrated circuit is the fact that if it becomes defective, it cannot be repaired. This is understandable; however, it also means that repairing a faulty circuit constructed of ICs is easy—just replace one IC (containing hundreds of components and costing less than a single transistor) instead of troubleshooting a complicated discrete circuit before the fault can be found. As you can see, the advantages far outweigh the disadvantages.

IC CONSTRUCTION

There are a number of different methods used in the manufacture of ICs. You are already familiar with some of these techniques from your study of the manufacture of older solid state devices such as diodes and transistors. Some terms you may not be familiar with in IC construction are the terms linear ICs and digital ICs. In practice, the structure of these devices are very nearly the same. The words linear and digital really describe modes of operation, but their constructions do differ slightly. This is because linear devices handle analog signals, those voltages and currents that vary at a linear rate, rather than voltages and currents changing abruptly as in a digital IC.

In contrast, digital devices must switch these voltages and currents, or transfer them as it is called, at considerably faster rates than linear ICs do. The transistors and diodes in the digital ICs then must be constructed slightly different for these high speeds to take place. In fact, digital ICs are grouped into families depending usually on the voltages used to power these devices and the speed at which these devices can transfer a signal from the input of the IC to its output.

Monolithic Construction

In monolithic construction, as the name implies, a single layer or substrate is used as the base for the construction of components that make up the complete IC. However, a single IC is never made. Instead, a thin semiconductor wafer (about 0.009 inches thick) is used as a substrate onto which hundreds of ICs are made. The wafer may be anywhere from 1 inch to 2 inches in diameter and is sectioned off into tiny little areas, usually rectangular in shape, called chips (Fig. 13-1). The manufacturing process in making an IC takes place simultaneously on hundreds of chips on a single wafer. Each chip will eventually contain all of the components necessary to make one complete IC.

Bipolar ICs

Usually a wafer of P-type material is used in making a bipolar

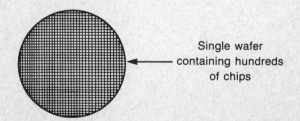

Single wafer
containing hundreds
of chips

Fig. 13-1. Each little square in this wafer will eventually become a complete IC in itself.

IC. (A bipolar IC contains transistors, resistors, diodes, and capacitors.) Figures 13-2 through 13-6 are an example of the processes involved in constructing an IC with two components, a diode and a transistor.

In Fig. 13-2, an N-type layer of semiconductor material is epitaxially grown onto a P-type substrate. This P-type substrate might be one small area of the chip or the entire chip. The epitaxial layer is usually from 10 to 20 micrometers in thickness. The silicon dioxide layer in Fig. 13-2 is due to exposing the epitaxial layer to steam at about 1750° F. This oxidizes on the N-type material, forming a very thin layer of silicon dioxide approximately 0.5 micrometers in thickness. The silicon dioxide layer is used as an insulating layer. Now the silicon dioxide layer is coated with a photoresist material. It is then covered with a mask that has selected areas removed, so that ultraviolet rays can pass through those areas and expose predetermined sections of the silicon dioxide. The mask is removed and the IC is given an acid bath that etches away the photoresist and parts of the silicon dioxide layer that have been exposed to the ultraviolet light.

This process, which leaves a pattern of openings in the silicon dioxide layer is called photolithography. Heavy concentrations of P-type impurities are then diffused through these window openings in the silicon dioxide layer forming P + semiconductor material. Those areas under the silicon dioxide layer not affected remain as N-type semiconductor material and are sometimes called isolation wells or islands, shown in Fig. 13-3.

The steam process is repeated, once again forming silicon dioxide over the entire surface. Windows are again formed through the process of photolithography and this time P-material is diffused into the N-material that was untouched in the first steps of

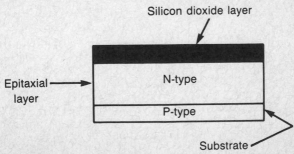

Fig. 13-2. The first step in the manufacture of a bipolar transistor and junction diode.

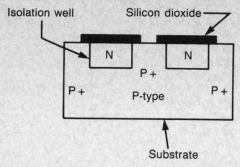

Fig. 13-3. The second step in the manufacturing process.

construction. As Fig. 13-4 shows, these P regions become the anode of a diode and the base of an npn transistor.

Once again steam deposits a silicon dioxide layer over the entire IC and through the process of photolithography and more diffusion, an N+ area forms in the P base and the P anode and N isolation wells of the transistor section of the IC, shown in Fig. 13-5.

Finally, aluminum or gold is vaporized through a metallization mask depositing interconnects and crossovers among the devices just formed, shown in Fig. 13-6. Figure 13-7 is another IC that has been constructed showing first the cross-section of the IC at the top, the connections between devices in the center, and the schematic of the IC at the bottom. As you can see, this IC contains all four of the basic components found in most of today's ICs: the capacitor, diode, transistor, and resistor.

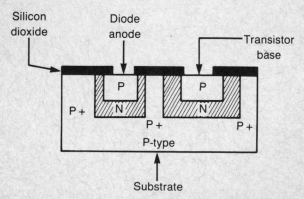

Fig. 13-4. Further along in the manufacturing process of the junction diode and bipolar transistor.

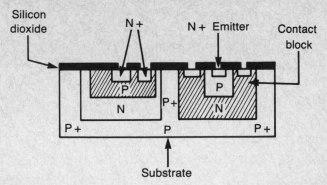

Fig. 13-5. The final step prior to constructing a working IC with a transistor and diode.

MOS IC Construction

You learned earlier in this study of solid state devices that bipolar transistors are only one of a variety of transistors that are used in electronic circuits. Other types include MOSFETs and different classifications of MOSFETs. Their construction is similar to that of the bipolar devices using a similar process of photolithography in a step-by-step basis in transistor construction.

In many of today's electronic circuits MOS devices are being selected over the bipolar type for transfer of information simply because MOS devices use less power, generate less heat, and are especially suited to high density (thousands of components) IC construction than bipolar types. In fact, in LSI (large scale

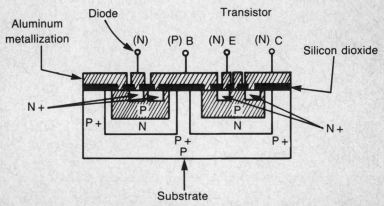

Fig. 13-6. The last step involves aluminum metallization and the attachment of leads.

293

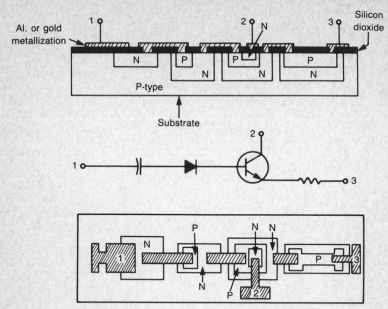

Fig. 13-7. Construction of an IC showing the comparisons among construction, schematic, and actual chip layout.

integration) and VSLI (very large scale integration) most standard methods used in the photolithographic process are simply not adaptable to these high density ICs. A new technique, still very much in its infancy, called electron lithography uses a beam of high energy electrons striking an electron resist material rather than ultraviolet rays striking a photo resist material. The electron beam is deflected in a modulated pattern, like the electron beam in a cathode ray tube and etching is done by ionized gases called plasmas instead of by acid baths. The accuracy of reducing an IC mask pattern down to one thousandth of its original size is significantly enhanced using this new type of technology rather than optical lenses, even those that have been designed using a computer.

Figure 13-8 is a simplified illustration of the construction of a P channel enhancement mode MOSFET. Here, P + material is diffused into an N-type substrate. The P + areas form contacts for the drain and source of the device. As you can see, silicon dioxide covers all of the top layer, except for a small area of P + material, and the silicon dioxide is very thin over the area between the drain and the source. Through a metallization mask, aluminum is deposited over the silicon dioxide. Part of this aluminum is the gate of the MOSFET and is separated from the substrate by the thin

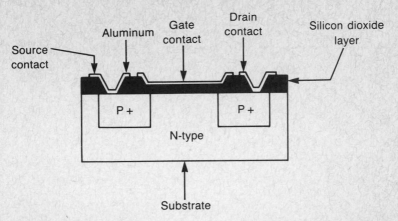

Fig. 13-8. Simple illustration showing the construction of a P channel enhancement mode MOSFET.

silicon dioxide layer, which is usually no more than 1000 angstroms thick. To operate in the enhancement mode, current flows from the source to the drain through a channel under the gate that is formed when the device is properly biased. Remember, this requires a negative bias supply voltage, V_{GS}, from gate to source.

Thin-Film Techniques

A thin-film IC uses a combination of techniques to produce the finished products. Thin-film ICs are used to produce resistors and capacitors while the standard monolithic technique is used to produce the diodes and transistors that are then added to the resistors and capacitors of the thin-film IC. A hybrid of sorts is developed using the thin-film method because finished components in chip form are added to complete the IC circuit.

The substrate of the thin-film IC is usually glass or ceramic. The resistors and capacitors are deposited onto the substrate as very thin layers of oxides and metals. Eventually diodes and transistors are added and are connected to the resistors and capacitors using extremely thin wires.

To produce a thin-film capacitor, a layer of metal is deposited onto a substrate, followed by a thin dielectric layer, and finally another thin layer of metal is deposited over the oxide dielectric to form the top plate of the capacitor.

To produce a thin-film resistor, tantalum or nichrome is deposited on the surface of the substrate in the form of a thin film or thin strip. The length, width, and thickness of the strip determine the value of the resistance. To interconnect these thin-film (0.1 thousandths of an inch thick) resistors, thin metal strips, deposited on the substrate made of gold, aluminum, or platinum, are used.

The depositing of these resistors and capacitors onto the substrate may involve two methods. One is the evaporation method where the material used to construct the component is heated within a vacuum until it evaporates and condenses onto the substrate. Usually a mask is used to make sure that the material is deposited in the right location. The second method is called the sputtering process and occurs in a gas-filled chamber. Here, the gas is ionized and bombarded with ions. The displaced atoms within the material are then deposited onto the substrate, again through a mask that allows only certain areas on the substrate to be coated.

Thick-Film Techniques

The similarity between thin-film and thick-film ICs lies in the fact that the resistors and capacitors are formed onto the substrate while transistors and diodes are constructed separately and then added later resulting in a complete IC circuit. This is where the similarity ends, because there are a number of major differences between the two.

In the thick-film IC, the device is, as the name implies, thicker than its cousin previously discussed. In fact, higher value capacitors must be constructed separately from the IC and then added at a later time during construction like the transistors and diodes. Adding these extra components gives a discernable difference in appearance, as well, between the two types of ICs. The thin-film IC looks as though the components have been painted onto the substrate surface, but this is not the case with the thick-film IC. In this respect, the thick-film IC construction method yields a device that is normally greater than 0.0001 inches in thickness.

Placing components, such as resistors, on a thick-film substrate involves a process called silk-screening. A screen is placed on top of the substrate in which specific designated areas of the screen are open while the remaining areas of this very thin screen (made from wire, not silk) are blocked off. The open areas form a pattern of resistors and capacitors needed to make the IC. Both the conductors (interconnects) and components are forced through the screen and then hardened through a heating process. The resistors and capacitors are usually laid down first, then the conductors are added, connecting the components together. Some areas are left open for diodes and transistors and for high value capacitors.

THE HYBRID

A hybrid is a combination of ICs, thin and thick-film resistors and capacitors, and discrete components such as diodes and transistors, in IC form. The hybrid itself contains a substrate for mounting these devices, sometimes ceramic, a hermetically sealed cover, and a system of pins for mounting in a PC board just like a typical IC. Figure 13-9 is a photo of a typical hybrid with its cover removed.

This particular hybrid contains two complex ICs, tunable mechanical filters, and even a crystal. It also contains 48 pins for mounting, similar to an ordinary IC, and a cover to enclose all of these components. This hybrid also uses a thick-film technique for interconnects. Probably the biggest advantage of hybrids is the complexity of circuit configurations that can be obtained by the mixture of different types of IC construction methods used to form an IC. However, hybrids are usually heavier and larger than the monolithic IC and become less reliable with the addition of discrete components.

Fig. 13-9. A typical hybrid containing a thin-film substrate and discrete components (John Sedor Photography).

IC PACKAGING AND NUMBERING

The packaging of an IC is similar to any other type of solid state device. In effect, the chip must be protected from the environment and is therefore hermetically sealed. The package itself is usually formed from plastic or epoxy using an injection molding machine, but some ICs used for military applications are packaged in ceramic. These are nonmetallic and can withstand temperatures considerably higher than the plastic types.

The chip itself sits in the center of the package, whether plastic or ceramic, connected to the IC's mounting pins with thin metal strips. The outside style of the package can be formed in a number of different styles, but the most popular is called the dual in-line package (DIP). This type of package is shown in Fig. 13-10. These ICs are 40-pin DIPs, but this type of packaging comes in 6-pin, 8-pin, 14-pin, 16-pin, and upwards, depending on the complexity of the IC or the number of ICs of the same type within the one package. Indeed, some packages contain 2, 4, or more identical circuits on one chip.

Another type of package is shown in Fig. 13-11. This IC is a voltage regulator and is packaged in a metal enclosure which can be mounted on a metal chassis which is used as a heat sink to draw heat away from the chip inside the can. If you look closely, you'll see that this is actually a hybrid mounted in this package. This particular device is a PIC625, a power switching regulator manufactured by Unitrode Corporation. It is used as the switching device in a switching power supply and is capable of handling a

Fig. 13-10. Dual in-line IC packages manufactured by three different companies (John Sedor Photography).

Fig. 13-11. A voltage regulator mounted in a power transistor type of package is this PIC625 switching regulator (courtesy Unitrode Corporation).

peak output current of 15 amperes; thus the provision for chassis mounting.

With all of these different types of packages and with the same IC sometimes found in different types of packages, its important for the technician to understand how ICs are identified. Although the industry has not established standards for all ICs, numbers pertaining to a particular IC from one manufacturer are similar to the numbers from another manufacturer making the same IC.

The identification number on an IC is normally split into five segments. As an example, look at the following number printed on an IC package:

SN74LS109J

SN 74 LS 109 J

The first part, SN, signifies that it is a bipolar device manufactured by Texas Instruments. 74 indicates a TTL family of devices, LS means low power Schottky, and 109 says that the device is a dual JK positive-edge triggered flip-flop. Finally, the J suffix tells the user that it is a high temperature ceramic package.

The number CD4032B can also be separated into five segments and identified as follows:

CD 40 XX 32 B

The CD signifies that the device is manufactured by RCA and is a digital IC circuit. 40 shows that it is a CMOS IC and the two X's show there is no subfamily. 32 indicates that the IC circuit within the package is a triple serial adder (three separate circuits instead

299

of just one) and B identifies this particular device as an RCA high voltage CMOS IC.

Prefixes on ICs usually signify who the manufacturer is, such as CD for RCA, DM or DS for National Semiconductor, UA for Fairchild Semiconductors, and UC for Unitrode Corporation. There are, of course, many, many more.

APPLICATIONS

Earlier it was stated that there are basically two kinds of integrated circuits. These are linear, or analog, ICs and digital ICs. A linear IC is a device that produces an output signal that is proportional to the input signal. In effect, the output signal is an exact replica of the input signal. If the output signal is larger in amplitude than the input signal, then the linear device is an amplifier because gain has been introduced. If the output signal amplitude is the same as the input signal amplitude then the gain of the linear IC is unity.

With linear ICs, the output is analogous (the same as) the input signal and they are therefore sometimes called analog ICs. Linear ICs can be found in many types of electronic circuits including communications, power supplies, and especially in home entertainment systems.

Digital ICs usually are made to produce one of two outputs, either zero volts (low) or +5 Vdc to +15 Vdc (high). These ICs perform a switching action or gating action and thus, many of the earlier, less complex digital ICs produced were called gates. Digital ICs, particularly gates, have one or more inputs and usually only one output. When there is more than one input, the digital IC is usually acting as a conditional device meaning that the output (high or low) is a condition of the input signal levels, whether they are high or low.

These types of digital ICs have evolved to some very complicated and highly advanced digital devices such as ROM (Read Only Memory), RAM (Random Access Memory or more appropriately, Read And Write-Memory), and microprocessors.

Linear Integrated Circuits

A very popular type of linear integrated circuit is the operational amplifier you read about in Chapter 9. It certainly fits the description of a linear IC. It amplifies and reproduces the input signal and can be used in many different ways, even as a filter or

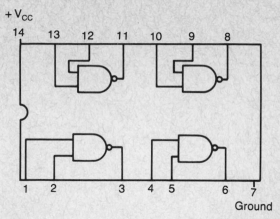

Fig. 13-12. A schematic representation of the IC of a digital IC called a quad two input NAND gate.

as an oscillator. In those ways it is acting like an analog device.

The operational amplifier, in integrated circuit form, is actually made up of many subcircuits consisting of transistors, diodes, resistors, and capacitors, but it is still one circuit. Some linear integrated circuits contain dozens of circuits that may, for instance, control the entire audio section of a color television receiver. Some ICs even contain most of the circuitry found in an AM or FM radio receiver. No doubt you've seen the slim style pocket radios made by some manufacturers of home electronic entertainment systems.

Digital Integrated Circuits

Most digital ICs are gates or are made up of gates that can perform much more complex tasks. Figure 13-12 is an example of the inside of a digital IC. This particular IC is called a quad two-input NAND gate and carries the identification number XX74XX00. Usually its labeled SN7400 or SN5400. This particular IC has four

Fig. 13-13. The truth table for the circuit of Fig. 13-12.

A	B	Y
0	0	1
0	1	1
1	0	1
1	1	0

Fig. 13-14. This type of IC, called an EPROM, is used in computer memory applications. (John Sedor Photography.)

identical types of digital circuits inside of it. Each circuit is called a NAND gate. When both inputs to a NAND gate are high, the output is low. (Actually the output would also be high when both inputs are high, but the out of this particular gate is inverted.)

All digital IC gates have a guide or table that explains what happens to the output when certain conditions are present on the input. These are called Truth Tables and the Truth Table for the NAND gate is shown in Fig. 13-13. What is obvious here is that a high output is seen in every case except when both inputs are high.

Digital gates are combined with each other to produce some very complex circuits, some of which are used in computer applications. Figure 13-14 is a photo of a computer IC called a memory IC. This IC is more technically termed an EPROM (an acronym for Erasable Programmable Read Only Memory IC). This IC can be programmed by writing a series of instructions into it called a program. This program gives detailed instructions to a microprocessor which initiates a series of actions in an electronic circuit. If the program is to be changed and a new program written into the EPROM, the device is subjected to an ultraviolet light which shines on the chip inside the IC through the glass window in the top of the package. After about 30 minutes the old program is erased and a new program can be written into the device.

As you can see, linear and digital ICs are used in a number of different ways and have experienced a tremendous growth in the past few years. This has been due in part to the computer revolution that has demanded that new ICs be developed to satisfy the needs of researchers and design engineers alike. With newer and more precise manufacturing techniques virtually what will be seen ahead in the IC industry, and in the electronics industry itself, will be limited only by what the human mind can imagine.

Lab Experiments

In most areas of electronics, hands on experience is essential so that you can better understand the real workings of the circuits you have studied. It is, of course, necessary to understand the academics of electronics, but the real challenge and final satisfaction in seeing the correlation between theory and practice comes with actually putting components together and seeing a circuit work. These experiments help you to see that correlation more clearly and can be modified as your understanding of solid-state devices increases.

As with any type of electronics projects there is a basic requirement in circuitry to get started. Each experiment has its own set of parts but there are components needed that are common to most of these experiments.

I have found that it's simply more convenient to have on hand a package containing the essential power sources, variable resistances, and breadboard assembly on which to construct the circuit. Also essential is a VOM. I prefer using a digital meter because the readout doesn't require any interpolation on my part and the result is usually carried out to three decimal points. Also, the input impedance of the DVM is such that loading down of the circuit is not a problem. Finally, various lengths of solid strand #22 gauge wire are necessary for making connections between components and for supply voltages and ground from the power sources. Figure E-1 is an example of the type of equipment recommended for these experiments.

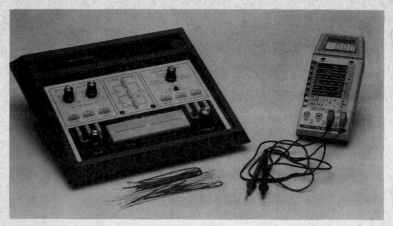

Fig. E-1. An alternative to separate equipment in performing experiments is by using a trainer like this one, manufactured by Heathkit, and a DVM shown here distributed by RSR Electronics (John Sedor Photography).

The lab board contains a variable dc voltage source proving zero to + or − 15 Vdc at 100 mA, an ac source providing both 15 Vac and 30 Vac at 60 Hz, an audio generator providing frequencies from 200 Hz to 20 kHz, and a breadboard for constructing circuits. Also included are two potentiometers of 1 k ohm and 100 k ohms.

The digital voltmeter has an LCD readout, high input impedance, and precision for measuring ohms, volts, and current down to the microampere range.

These experiments can also be performed using power sources and other components that are not neatly packaged, but the convenience certainly outweighs the disadvantages of separate items.

EXPERIMENT 1:   TESTING THE PN JUNCTION DIODE

The purpose of testing most diodes is simply to determine whether or not the diode is good or bad, called go/no go testing. Testing for manufacturer's specifications is usually not required. An ohmmeter can be used for this type of good/bad test method. This experiment is designed to show you how to test most standard PN junction diodes.

Required Materials

1 - Experimenter's board as shown in Fig. E-1 or equivalent

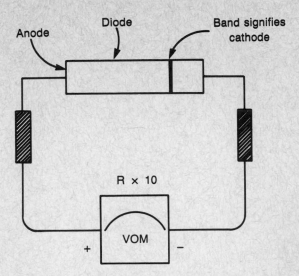

Fig. E-2. Measuring the forward resistance of a diode.

1 - Silicon diode, 1N4001 or equivalent
1 - VOM or DVM

Procedure

1. Set the ohmmeter to the lowest range and connect the VOM to the diode as shown in Fig. E-2. This measures the forward resistance of the diode. Record this value in Table 1 under R_F.

2. Set the ohmmeter to the RX100K range and connect the meter to the diode as shown in Fig. E-3. This measures the reverse resistance of the diode. Record your value in Table E-1 under R_R.

3. Now calculate the ratio of reverse resistance to forward resistance and record this ratio in Table E-1 under R_R/R_F.

Discussion

This experiment shows that the forward resistance of the diode is very low in comparison to the reverse resistance of that same diode. In fact, you should have seen a ratio of at least 10:1. Usually

Table E-1.

R_F	R_R	$R_R{:}R_F$

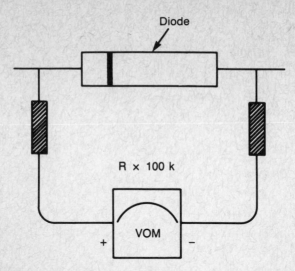

Fig. E-3. Measuring the reverse resistance of a diode.

this is much higher. Any ratio lower than 10:1 usually indicates a bad diode.

It is a good idea to try this experiment with other diodes such as those made of germanium just to see the difference in resistances between forward and reverse measurements in that type of semiconductor material.

EXPERIMENT 2: THE ZENER DIODE

You saw in Chapter 3 that the zener diode is very popular as a device for regulating voltage. In fact, it is the basis for most voltage regulators, whether as the regulator itself, or as a reference device within the overall scheme of a more complex voltage regulator such as a switching regulator using an op amp. This experiment shows you the ability of a zener diode to regulate voltage.

Required Materials
1 - Experimenter's board as shown in Fig. E-1 or equivalent
1 - VOM or DVM
1 - Zener diode, 1N743 or equivalent
1 - 47 ohm, 1/2 watt resistor
1 - 100 ohm, 1/2 watt resistor
1 - 220 ohm, 1/2 watt resistor
1 - 470 ohm, 1/2 watt resistor
1 - 1 k ohm, 1/2 watt resistor

Procedure

1. Construct the circuit shown in Fig. E-4. Notice that the zener diode is connected so that it is reverse biased in this circuit.

2. Adjust the input voltage to 9 Vdc by measuring the voltage from the anode of the zener diode to the arm of the potentiometer while adjusting R_V until you read the required 9 Vdc.

3. With no load on the output of the regulator, measure the voltage across Z_1 and record your results in Table E-2 under output voltage.

4. Next, connect a 1 k ohm resistor as the load across Z_1. Table E-2 shows that this load resistor will draw about 5 mA of current. Measure the voltage across Z_1 and record your results as you did in step 3.

5. Repeat step 4, this time using load resistors as specified in Table E-2. Then record the voltage across Z_1 in Table E-2 for each load resistor.

Discussion

You can see that the zener diode of this circuit is capable of regulating (maintaining) the output voltage of this circuit under varying load conditions. The regulator, however, should have decreased significantly when the 47 ohm resistor was used as the load. This is because with a 9 Vdc input voltage, the maximum load current of 39 mA was exceeded, effectively overloading the zener regulator, causing the voltage to drop below the 5.1 volt level of the zener. This 39 mA comes from the fact that with a 9 Vdc input, 5.1 volts is dropped across the zener diode while 3.9 volts is dropped across R_S. Thus, according to Ohm's law, 3.9 volts dropped across a 100 ohm resistor causes 39 mA to flow.

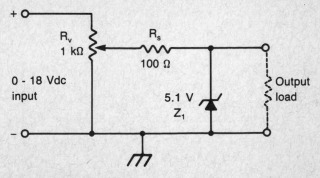

Fig. E-4. Zener diode voltage regulator.

R$_L$ (ohms)	V$_{out}$	I$_L$ (mA)
No load		0
1000		5
470		11
220		23
47		108

EXPERIMENT 3: THE BIPOLAR TRANSISTOR

This experiment is designed to give you a hands on idea of how a typical bipolar transistor is tested. Again, as with the PN junction diode, this is a good/bad, go/no go, type of measurement. Both NPN and PNP transistors can be measured using this technique, but only the NPN transistor is shown.

Required Materials

1 - Experimenter's board as shown in Fig. E-1 or equivalent
1 - VOM or DVM
1 - NPN transistor, 2N2222 or equivalent

Procedure

1. Place the transistor on a flat work surface so that the flat side of the device is facing up. If the transistor is not marked, locate the emitter, base, and collector leads for this particular device by reading from left to right. The emitter is the lead on the left, the base is the middle lead, and the collector is the right lead. To measure the forward resistance of the emitter-base junction and collector-base junction use the R × 10 or R × 100 range of the VOM.

2. Connect the positive lead of the VOM to the base lead and the negative lead of the VOM to the emitter lead, shown in Fig. E-5. Record your reading in Table E-3 under Forward Res., E-B.

3. Now measure the forward resistance of the collector-base junction by connecting the positive lead to the base and the negative lead to the collector as shown in Fig. E-5. Record your reading in Table E-3 under Forward Res., C-B.

4. Next, reverse the leads from step two and measure the reverse resistance between emitter and base and between collector and base

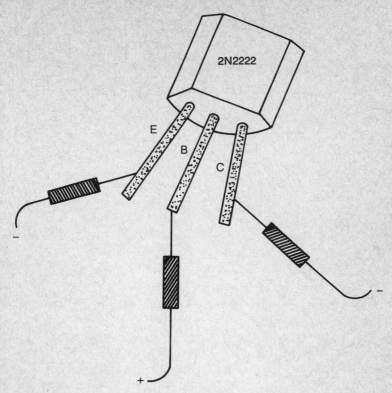

Fig. E-5. Testing the bipolar junction transistor.

and record your results in Table E-3 under Reverse Res., E-C and C-B, respectively.

Discussion

The forward resistance of each junction should have been very low, just like the ordinary PN junction diode. However, the reverse resistance should have shown an almost infinite resistance reading for either junction. An important point here is that the readings in the forward direction and in the reverse direction should be significantly different. The PNP bipolar transistor is tested the same

	Forward res.		Reverse res.	
Table E-3.	E-B	C-B	E-B	C-B

way, but keep in mind that for each step in testing the leads are reversed.

EXPERIMENT 4: THE IGFET

The purpose of this experiment is to familiarize you with the characteristics of the insulated gate field effect transistor, IGFET. This is to help you to better understand the relationship between drain-to-source voltage, V_{DS}, gate-to-source voltage, V_{GS}, and drain current, ID.

Required Materials

1 - Experimenter's board as shown in Fig. E-1 or equivalent
1 - VOM or DVM
1 - N Channel depletion mode IGFET (MOSFET), ECG220 or equivalent
1 - 1 k ohm, 1/2 watt resistor
1 - 4.7 k ohm, 1/2 watt resistor
1 - 10 k ohm, 1/2 watt resistor
1 - 100 k ohm potentiometer
1 - 1 k ohm potentiometer

Procedure

1. Construct the circuit of Fig. E-6.
2. Make sure that the leads of the IGFET are shorted with a thin

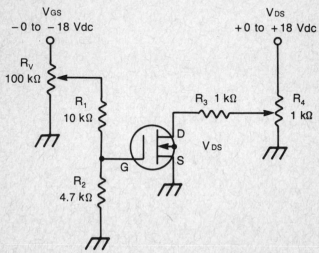

Fig. E-6. IGFET biasing circuit.

wire. Keep them that way until the circuit is completely constructed.

3. Drain current will be measured by measuring the voltage drop across R_3 and using Ohm's law to calculate I_D.

4. Adjust R_{V_1} fully counterclockwise.

5. Turn on power and adjust R_{V_2} until VDS is equal to 0.5 volts.

6. Measure the voltage drop across R_3. Calculate I_D and record this value in Table E-4 under 0.5.

7. Repeat step 5 for each value of VDS in Table E-4 and record the values of I_D where indicated for each value of VDS.

8. Now adjust R_{V_1} until VGS equals -0.5 volts.

9. Again, adjust R_{V_2} until VDS equals 0.5 volts.

10. Measure the voltage across R_3 and calculate I_D according to Ohm's law. Record your result in Table E-5 in the appropriate box.

11. Repeat steps 9 and 10 again for each value of VDS listed in Table E-5 while leaving VGS set at -0.5 volts.

12. Turn off the supply voltages and reverse the negative VGS supply, making it a positive VGS supply.

13. Now adjust VGS for $+0.5$ volts and VDS for $+0.5$ volts.

14. Measure the voltage across R_3 and calculate I_D according to Ohm's law. Record your result in Table E-6 in the appropriate block.

15. Continue adjusting VDS for the values shown in Table E-6 while recording the corresponding values of I_D.

16. Finally, plot the drain current versus drain-to-source voltage, for each value of gate-to-source voltage in Fig. E-7.

Discussion

As you can see, the IGFET can operate with either positive or negative gate voltages. Drain current does indeed flow when VGS equals zero volts, and drain current also flows when a negative VGS is applied, but most drain current flows when there is a positive gate-to-source voltage.

EXPERIMENT 5: THE PHOTOTRANSISTOR

The phototransistor in this experiment is the same as that discussed in Chapter 6. This experiment shows the proper biasing technique for the phototransistor and its typical optical characteristics.

Table E-4.

	VGS = 0 volts				
VDS (volts)	0.5	1	2	3	4
I_D (mA)					

Table E-5.

VGS = −0.5V					
VDS (volts)	0.5	1	2	3	4
I_D (mA)					

Required Materials

1 - Experimenter's board as shown in Fig. E-1 or equivalent
1 - VOM or DVM
1 - Phototransistor, 417-851 or equivalent
1 - 12 Vdc lamp
1 - 10 k ohm, 1/2 watt resistor
1 - 1 k ohm potentiometer

Procedure

1. Construct the circuit shown in Fig. E-8. Place the lamp next to the window of the phototransistor, approximately 1/8″ to 1/2″ away.

2. Using the VOM, adjust the positive supply voltage for +15 Vdc.

3. Using the VOM, adjust the negative supply voltage for −12 Vdc.

4. Cover the window of the phototransistor so that no light can get to the base of the device. Electrical tape works well for this step. Measure the voltage across R_C and record your result in Table E-7 under VRC with no light.

5. Next, calculate the current through R_C and record that in Table E-7 under I_{R_C} with no light.

6. Now, remove the cover from the window of the phototransistor and, while adjusting R_V until the lamp is at its brightest, measure the voltage once again across R_C. Record this voltage across R_C in Table E-7 under VRC with maximum light.

Table E-6.

VGS = +0.5V					
VDS (volts)	0.5	1	2	3	4
I_D (mA)					

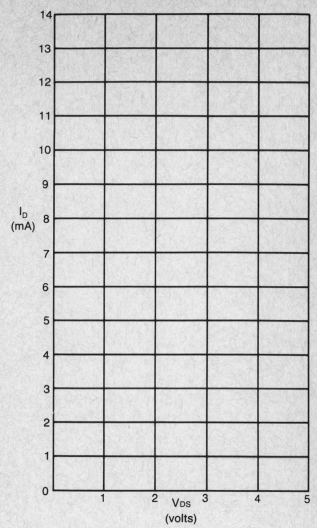

Fig. E-7. Vᴅꜱ versus I_D for the IGFET of Fig. E-6.

7. Finally, calculate the current through R_C and record this in Table E-7 also.

Discussion

When you calculated the current through the phototransistor and the window was covered, you should have calculated a very low current, called the dark current. The lower the dark current, the better the phototransistor.

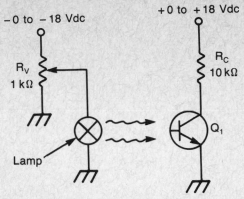

Fig. E-8. Circuit for testing the optical characteristics of a phototransistor.

When you removed the cover, you noticed the voltage across R_C increased as the lamp intensity increased. As a result, the current through R_C also increased. As you can see, the phototransistor controlled the current and voltage through the load and is a function of the intensity of the light.

EXPERIMENT 6:
BIPOLAR TRANSISTOR CHARACTERISTICS

This experiment is designed to show you the relationship among base and collector currents, and V$_{CE}$ or collector-to-emitter voltage, in a common emitter circuit configuration. All of these parameters can best be viewed in terms of the transistor's V-I characteristic curves. Therefore, the V-I curve will be plotted using the results obtained from the measurements taken in this experiment.

Required Materials

1 - Experimenter's board as shown in Fig. E-1 or equivalent
1 - VOM or DVM
1 - NPN transistor, 2N2222 or equivalent
1 - 1 k ohm, 1/2 watt resistor

No light		Maximum light	
V_{RC}	I_{RC}	V_{RC}	I_{RC}

Table E-7.

314

1 - 10 k ohm, 1/2 watt resistor
1 - 100 k ohm potentiometer
1 - 1 k ohm potentiometer

Procedure

1. Construct the circuit shown in Fig. E-9.
2. Adjust R_{V_1} fully counterclockwise and R_{V_2} to midrange.
3. Turn on the power and adjust VCC for + 15 Vdc.
4. Adjust I_B to 10 μA. This is done by connecting a voltmeter across R_B and slowly adjusting R_{V_1} until a voltage drop of 0.1 volts is achieved.
5. Now adjust R_{V_2} until VCE equals 1 volt.
6. Calculate the current through R_C by measuring the voltage drop across it and using Ohm's law. This is collector current, I_C. Record your calculated result in Table E-8 in the first block under 1.
7. Complete Table E-8 by adjusting VCE to the indicated voltages and recording your calculated I_C values.
8. Once again, while measuring the voltage drop across R_B, adjust R_{V_1} for an I_B of 20 μA. This occurs when the voltmeter reads 0.2 volts.
9. Now adjust R_{V_2} until VCE equals 1 volt.
10. Now measure the voltage drop across R_C and calculate I_C. Record this calculated value in Table E-9 in the appropriate block.
11. Continue to adjust R_{V_2} for the indicated values in Table E-9 and record the calculated values of I_C in the appropriate blocks.

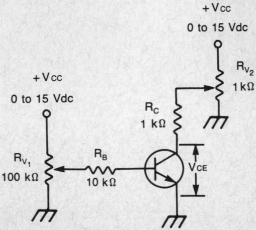

Fig. E-9. Test circuit for establishing V-I characteristic curve of a bipolar junction transistor.

$(I_B = 10\mu A)$

VCE (volts)	1	2	3	4	5	6
I_c (mA)						

Table E-8.

12. Now adjust R_{V_1} until the voltage drop across R_B is 0.3 volts. This will produce a base current of 30 μA.

13. Next, adjust R_{V_2} for a VCE of 1 volt.

14. Measure the voltage drop across R_C and calculate I_C from Ohm's law. Record your calculated value in Table E-10.

15. Repeat step 13 for each value of VCE listed in Table E-10 and record your values of I_C in the same table.

16. Graph I_C versus VCE for the values in Table E-8 onto Fig. E-10. Connect the points plotted and label the curve $I_B = 10$ μA.

17. Do the same for Tables E-9 and E-10 and label those curves $I_B = 20$ μA and $I_B = 30$ μA respectively.

Discussion

As you can see, these characteristic curves do not show the

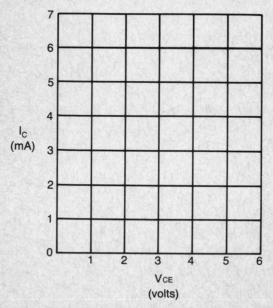

Fig. E-10. V-I characteristic curve of the circuit of Fig. E-9.

$$(I_B = 20\mu A)$$

Vce (volts)	1	2	3	4	5	6
I_c (mA)						

Table E-9.

initial rise in I_C since the Vce values began at 1 volt instead of zero volts for each value of I_B. However, there is sufficient data here to show the gain of the transistor. By selecting a Vce of 4 volts or more and determining the change in collector current for a given change in base current, the dc beta, β, of the device can be found. The alpha can also be found simply by first finding beta and then using the following equation:

$$\alpha = \frac{\beta}{\beta + 1}$$

EXPERIMENT 7: THE OPERATIONAL AMPLIFIER

One of the more interesting devices is the operational amplifier IC. I can think of no less than three dozen ways in which to use op amps, and that does not include all of the ways in which op amps can be configured. This experiment deals with the op amp as it is used as a window detector. This circuit gives a visual indication of when a particular voltage is within a narrow minimum and maximum range of voltages or within a window of voltages. Since the upper and lower limits are framed, this circuit is called a window detector. The circuit indicates when the monitored voltage is within the window by acting as an alarm to trigger one of two LEDs when the voltage is above or below that set window.

Required Materials

1 - Experimenter's board as shown in Fig. E-1 or equivalent
1 - VOM or DVM

$$(I_B = 30\mu A)$$

Vce (volts)	1	2	3	4	5	6
I_c (mA)						

Table E-10.

1 - ECG834 IC op amp comparator or equivalent
2 - 1 k ohm, 1/4 watt resistors
1 - 100 k ohm potentiometer
1 - 1 k ohm potentiometer
2 - LEDs, 412-40 or equivalent

Procedure

1. Construct the circuit shown in Fig. E-11. Make sure the LEDs are in the circuit properly.

2. Apply a 5.0 Vdc signal to V_{in} and notice the status of the LEDs. A digital voltmeter is better for setting this input voltage since it must be within 5 percent of the 5 Vdc required. Record in Table E-11 which LED, if any, is on and which is off.

3. Apply a 4.0 Vdc signal to V_{in} and notice the status of the two LEDs. Record in Table E-11 which LED is on and which LED is off.

4. Apply a 6 Vdc signal to V_{in} and notice the status of the two LEDs. Record in Table E-11 which LED is on and which LED is off.

Discussion

As you can see from step 2 in the procedure, a 5 Vdc input signal is centered in value between the upper limit of 5.5 Vdc and the lower limit of 4.5 Vdc of this voltage window detector.

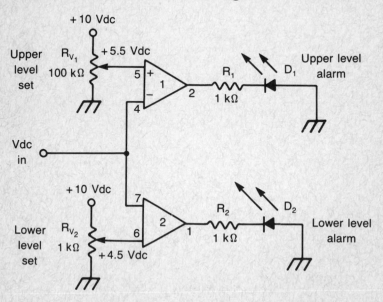

Fig. E-11. Op amp comparators used as a high-low level alarm circuit.

VDC$_{in}$	LED1	LED2
5		
4		
6		

Table E-11.

Therefore, neither LED should be on with an input signal voltage of 5.0 Vdc. Once the Vdc input level falls above or below the upper or lower limits, the appropriate LED lights. A circuit such as this LED alarm circuit can be very useful where a limit of 10 percent deviation is required of a 5 volt supply used in a critical circuit such as a computer circuit.

EXPERIMENT 8: THE UJT RELAXATION OSCILLATOR

The UJT is a thyristor that can be configured as an oscillator to give two different types of output waveforms, a sawtooth and a pulse. This experiment is set up so that you can see the oscillator functioning by viewing the output of Base 1 with an LED.

Required Materials

1 - Experimenter's board as shown in Fig. E-1 or equivalent
1 - UJT, ECG6400B or equivalent
1 - 47 ohm, 1/4 watt resistor
1 - 100 ohm, 1/4 watt resistor
1 - 100 k ohm, 1/4 watt resistor
1 - 33 k ohm, 1/4 watt resistor
1 - 35 μF electrolytic capacitor
1 - LED, 412-40 or equivalent
1 - VOM or DVM

Procedure

1. Take the unijunction transistor and lay it down on a work surface with the flat side up. The leads are then, from left to right, Base 1, emitter, and Base 2.
2. Construct the circuit shown in Fig. E-12. Use a 100 k ohm resistor for R$_x$.
3. When the power is turned on, notice the length of time between pulses on the LED.
4. Now turn off the power and replace the 100 k ohm resistor with a 33 k ohm resistor.

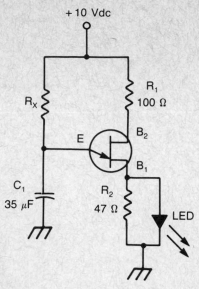

Fig. E-12. A relaxation oscillator using a unijunction transistor as the active device.

5. Turn on the power and again measure the length of time from one LED flash to the next.

Discussion

This is a very simple relaxation oscillator. Using the LED to see if oscillations are taking place is, of course, considerably cheaper than using an oscilloscope. You should have noticed that the length of time between flashes using the 100 k ohm resistor is about 3.5 seconds as compared to about 1.1 second using the 33 k ohm resistor. You could also have changed the pulse rate by changing the size of the capacitor, because the frequency of oscillations is a function of both the capacitor and the emitter resistor of the UJT. Also, if you take your VOM and place it in parallel with C_1, you'll see the capacitor charge up and then quickly discharge each time an oscillation takes place.

EXPERIMENT 9:   THE 555 TIMER

The 555 timer is a very versatile IC and can be used in a number of timing applications, either triggered or free-running. It is also possible to configure this timing device as a pulse width modulator. This experiment is a very simple circuit consisting of only a few

320

external components configuring the 555 timer as an astable, or free-running, pulse generator. As with the previous experiment, an LED is used so that you can see the oscillations occurring and measure them with a stop watch.

Required Materials

1 - Experimenter's board as shown in Fig. E-1 or equivalent
1 - 555 IC
1 - 560 ohm, 1/4 watt resistor
1 - 100 k ohm, 1/4 watt resistor
1 - 0.01 µF capacitor
1 - 35 µF electrolytic capacitor
1 - LED, 412-40 or equivalent

Procedure

1. Construct the circuit shown in Fig. E-13.
2. Turn the power on. Using a stop watch, time the total length of time that the LED is both on and off. Start from when the LED turns on, continue when it turns off, then stop your timing as soon as it turns on again.

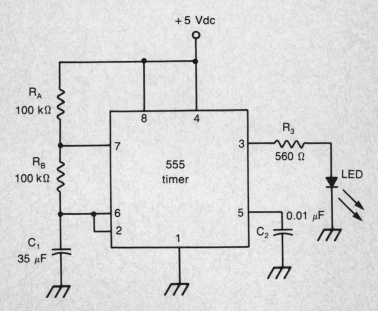

Fig. E-13. The 555 timer configured in the astable, or free-running, mode of operation.

3. The length of time that the LED remains on is a function of C_1, R_A, and R_B. This on time is when the output pulse from pin 3 of the I_C is high and the LED is on. It can be found from the following equation:

$$t_{high} = 0.695(R_A + R_B)C_1$$

Calculate this value and record your result here..........

4. The length of time that the LED remains off is a function of C_1 and R_B. This off time is when the output pulse from pin 3 of the IC is low and can be found from the equation:

$$t_{low} = 0.695 \, R_B C_1$$

Calculate this value and record your result here..........

5. The total on time and off time, or one complete pulse cycle, can be calculated from the equation:

$$T = 0.695(R_A + 2R_B)C_1$$

Calculate this value and record your result here..........

6. How does this calculated total time compare to the time on your stop watch?

Discussion

The astable multivibrator is a pulse circuit that can have many applications and is, indeed, one of the easier pulse circuits to build. The values for R_A, R_B, and C_1 in Fig. E-13 can be changed to obtain just about any timing sequence desired. For a 50 percent duty cycle where the time on equals the time off, the following equations are used:

$$t_{high} = 0.695 \, R_A C_1$$
$$t_{low} = 0.695 \, R_B C_1$$
$$T = 0.695(R_A + R_B)C_1$$

In these equations $R_A = R_B$ and a diode, connected with a cathode at pin 2 and anode at pin 7, is placed in parallel to R_B. The diode can be a standard 1N914 type.

322

EXPERIMENT 10: THE SWITCHING POWER SUPPLY

The switching power supply is used in many applications requiring high efficiency and low power dissipation. When the output voltage of the switching supply is equal to or greater than the reference voltage, no power is drawn from the unregulated power supply because the transistor switch is not driven on, thus adding to the switching power supplies efficiency.

Required Materials

1 - Experimenter's board as shown in Fig. E-1 or equivalent
1 - LM723 IC
1 - 1N914 diode
1 - 1 mH coil
1 - 100 μF electrolytic capacitor
1 - 0.1 μF capacitor
2 - 47 ohm, 1/4 watt resistors
1 - 100 ohm, 1/4 watt resistor
1 - 10 k ohm, 1/4 watt resistor
1 - 31 k ohm, 1/4 watt resistor
1 - 1 M ohm, 1/4 watt resistor
2 - 10 k ohm potentiometers
1 - PNP transistor, 2N3906 or equivalent
1 - VOM or DVM

Procedure

1. Construct the circuit shown in Fig. E-14.
2. Using your VOM adjust R_{V_2} for its maximum resistance.
3. Using your VOM adjust R_{V_1} for its maximum resistance.
4. Adjust your positive power supply for +12 Vdc and apply power to the circuit.
5. Connect your VOM to the output of the regulator and slowly adjust R_{V_1} until a reading of +5 Vdc is seen at the output.
6. An interesting method for viewing the regulation of this type of power supply is by plotting the output voltage, V_{out}, versus output current, I_{out}, under varying load conditions. R_{V_2} plus R_6 make up the load resistance. If you turn off the +12 Vdc to the circuit and adjust R_{V_2} in 50 ohm increments starting at 100 ohms you can plot the output characteristics of this switching regulator.
7. Begin by setting R_{V_2} to 100 ohms without power applied to

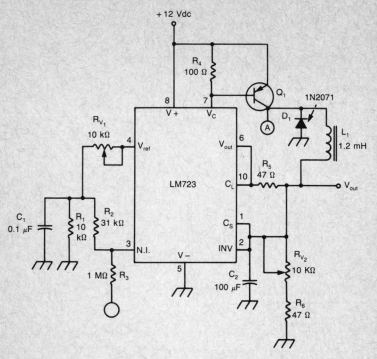

Fig. E-14. A typical 723 configured as a switching regulator.

the circuit. Then apply power and calculate output current, I_{out}, by using the equation:

$$I_{out} = V_{out}/(R_{V_2} + R_6)$$

To find V_{out} for this equation merely measure the output voltage with your VOM. Record your values for V_{out} and I_{out} in Table E-11 next to 100 ohms.

8. Repeat step 7, turning off the +12 Vdc supply to the circuit and adjusting R_{V_2} to 150 ohms. Again, turn the power back on and measure the output voltage of the regulator. Calculate the output current from the equation given in step 7 above. Continue to do this for each value of R_{V_2} given in Table E-11 until the table is complete.

9. Now go to Fig. E-15 and plot V_{out} versus I_{out} from the data gathered in Table E-11. You will then have the characteristic curve of this particular switching regulator.

Load $(R_{V2} + R_6)$	V_{out} (volts)	I_{out} (mA)
5 kΩ		
4.5 kΩ		
4.0 kΩ		
3.5 kΩ		
3.0 kΩ		
2.5 kΩ		
2.0 kΩ		
1.5 kΩ		
1.0 kΩ		
950 Ω		
900 Ω		
850 Ω		
800 Ω		
750 Ω		
700 Ω		
650 Ω		
600 Ω		
550 Ω		
500 Ω		
450 Ω		
400 Ω		
350 Ω		
300 Ω		
250 Ω		
200 Ω		
150 Ω		
100 Ω		

Discussion

The LM723 voltage regulator has been one of the more popular voltage regulators over the last several years. It can also be used as a voltage regulator to limit output current which is called a foldback power supply. There are now power supply ICs that are actually more diverse than the 723, but this has been the workhorse for many years and will probably continue to be so.

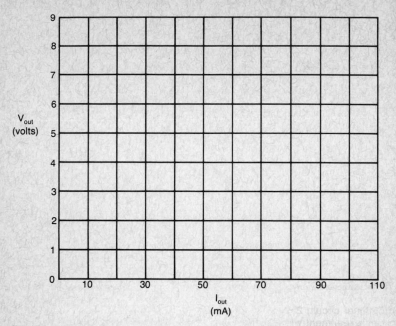

Fig. E-15. V-I characteristic curve for a 723 switching regulator.

Index